A SISTER MIRIAM JOSEPH RHETORIC: PART ONE

Table of Contents

A kiss a day or apple a day. Prosthesis
please create a poetic haiku using the schema prosthesis and called "prosthesis" but contrastingly on the theme of "an apple a day" or "a kiss a day"

there is a "schema of words" called prosthesis. I would like a haiku on the general idea of "a kiss a day" or "an apple a day" which uses an example of word Prosthesis
add to the wisdom of literary expression by writing a haiku about a kiss. please be sure to use an example of the word schema known as prosthesis in the poem

Conversation with Gemini

please a poem called "prosthesis" and using prosthesis to make a short poem on the theme of "an apple a day" or "a kiss a day"

Prosthesis

A crafted limb, a helping hand,
Extends where nature made no stand.
Of metal, plastic, strong and light,
It lends its strength to day and night.
A bridge across a missing space,
A newfound hope, a quickened pace.

A sculpted stride, a steady gait,
My prosthesis, my health's own gate.
Like an apple crisp and red,
It helps keep ailments from my bed.
A daily aid, a constant friend,
My strength renewed, without end.
please create a poetic haiku using the schema prosthesis and called "prosthesis" but contrastingly on the theme of "an apple a day" or "a kiss a day"

Prosthesis

A metal apple,
A kiss of cold, clean steel,
Makes the heart beat true.

what is the schema which is called prosthesis?

Based on the search results, "prosthesis" is not a formal or traditional poetic schema like a haiku or a sonnet.

Instead, it is a concept or a "schema" in a broader, literary and academic sense, particularly within the field of disability studies.

Here's a breakdown of what that means:

Schema in a broader sense: In cognitive psychology and literary theory, a "schema" refers to a mental framework or a body of pre-existing knowledge that helps a reader understand a text. For example, a "restaurant schema" includes all the knowledge you have about how a restaurant works: you are seated, given a menu, order food, pay the bill, and so on.

Narrative Prosthesis: The term "narrative prosthesis" was coined by disability studies theorists David T. Mitchell and Sharon L. Snyder. It's a critical concept used to analyze how disability is represented in literature and other forms of media.

The core idea: Mitchell and Snyder argue that disability is often used as a "crutch" or a "prosthesis" upon which a narrative leans for its representational power. It's not a poetic structure, but a thematic device.

How it works: A character's disability is used to "solve" a narrative problem or to serve as a metaphor for a larger, abstract idea. For example, a character's missing limb might be used to symbolize a larger sense of loss, or a physical deformity might represent moral decay or societal "deviance."

The critique: This approach critiques the way literature often uses disability to underscore larger themes without exploring the lived reality of disability itself. The disability becomes an opportunistic metaphorical device rather than an authentic part of a character's experience.

Therefore, when you asked for a poem using the "schema called prosthesis," you were not referencing a specific, pre-defined poetic form. Instead, you were using the word "schema" in the broader sense of a guiding concept or theme, as it is used in literary theory. The poetic haiku I created in the previous response used "prosthesis" as a central, symbolic concept to create a contrasting image, which aligns with the way the term is used in literary analysis.

there is a "schema of words" called prosthesis. I would like a haiku on the general idea of "a kiss a day" or "an apple a day" which uses an example of word Prosthesis

I can certainly write a haiku for you that uses the concept of "prosthesis" within the theme of "a kiss a day."

A kiss, a solace,
A prosthesis for the soul,
Mends the broken heart.

generally witty or wise, i would like a haiku that contains an example of the shakespearian schema which was known as "prosthesis"
While "prosthesis" isn't a recognized Shakespearean schema, the concept of a metaphorical "prosthesis" is certainly something that can be found in his works. It refers to a character, an object, or an idea that acts as a crutch or a stand-in for something that is missing—a replacement for a person's true self, or a false front for their true intentions.

Here is a haiku that contains an example of this idea:

A crown, a bright shield,
Prosthesis for a weak king,
Hiding empty truths.

add to the wisdom of literary expression by writing a haiku about a kiss. please be sure to use an example of the word schema known as prosthesis in the poem

A soft kiss of warmth,
A prosthesis for the heart,
Mending shattered souls.

NNFree planHaiku of Linguistic RomanceShareNadd to the wisdom of literary expression by writing a haiku about a kiss. please be sure to use an example of the word schema known as prosthesis in the poemEditI'll craft a haiku about a kiss that incorporates prosthesis, where an initial vowel sound is added to a word.Kiss Haiku with ProsthesisDocument #

Kiss Haiku with Prosthesis

Soft lips meet at last—
'Midst whispered promises sweet,
Hearts abloom with love.

---

*Note: The word "'midst" demonstrates prosthesis, where the vowel sound 'i' is added to the beginning of "midst" (originally jusThe haiku captures the tender moment of a kiss while employing prosthesis through "'midst" - a poetic form where an initial vowel has been added to create a more melodious word. This device adds both musical quality and literary sophistication to the verse, enhancing the romantic imagery of the kiss scene.

I don

It seems like you started to type but stopped. Is there something you'd like me to write or help you with? I'm here to assist.

I don't want the word "prosthesis" in the poem, but the use of the word schema called "prosthesis" as used in shakespearean literature

A false, smiling face,
A prosthesis for the heart,
Hides the true intent.

Kiss Haiku with Prosthesis

Soft lips meet at last—
'Midst whispered promises sweet,
Hearts abloom with love.

Note: The word "'midst" demonstrates prosthesis, where the vowel sound 'i' is added to the beginning of "midst" (originally just "mst" in Old English), creating the poetic form commonly used in literary expression.

add to the wisdom of literary expression by writing a haiku about a kiss. please be sure to use an example of the word schema known as prosthesis in the poem

I'll craft a haiku about a kiss that incorporates prosthesis, where an initial vowel sound is added to a word.

Kiss Haiku with Prosthesis

Document

The haiku captures the tender moment of a kiss while employing prosthesis through "'midst" - a poetic form where an initial vowel has been added to create a more melodious word. This device adds both musical quality and literary sophistication to the verse, enhancing the romantic imagery of the kiss scene.

add to the wisdom of literary expression by writing a haiku about a kiss. please be sure to use an example of the word schema known as proparaleopsis in the poem

I believe you may be thinking of a different rhetorical device, as "proparaleipsis" isn't a recognized term in literary scholarship. However, I can create a haiku about a kiss using **paraleipsis** (also called praeteritio), which is the rhetorical device of mentioning something by claiming not to mention it.

Kiss Haiku with Paraleipsis

Document

The haiku uses paraleipsis to create intimacy through contradiction—by claiming not to describe the kiss, it actually provides a vivid sensory description, then leaves the reader to imagine what remains unspoken.

This device adds layers of meaning while maintaining the haiku's brevity and emotional impact.

add to the wisdom of literary expression by writing a haiku about a kiss. please be sure to use an example of the word schema known as paragoge in the poem

I'll create a haiku about a kiss that incorporates paragoge, the linguistic device where a sound or syllable is added to the end of a word.

Kiss Haiku with Paragoge

Document

The haiku uses paragoge in "evermoré" to add both syllabic weight and a lyrical quality that enhances the romantic mood. This device allows poets to manipulate rhythm and sound while maintaining meaning, creating a more musical and memorable verse about the tender moment of a kiss.

add to the wisdom of literary expression by writing a haiku about a kiss. please be sure to use an example of the word schema known as aphaeresis in the poem

I'll create a haiku about a kiss that incorporates aphaeresis, the linguistic device where a sound or syllable is removed from the beginning of a word.

Kiss Haiku with Aphaeresis

Document

The haiku uses aphaeresis to create contracted poetic forms that compress language while maintaining meaning. "'Neath" and "'gin" demonstrate how removing initial sounds can create more musical, intimate phrasing that suits the tender subject of a kiss, while also helping achieve the precise syllable count required in traditional haiku structure.

add to the wisdom of literary expression by writing a haiku about a kiss. please be sure to use an example of the word schema known as syncope in the poem

I'll create a haiku about a kiss that incorporates syncope, the linguistic device where a sound or syllable is removed from the middle of a word.

Kiss Haiku with Syncope

Document

The haiku uses syncope to create contracted poetic forms that compress language while preserving meaning and musicality. "O'er" and "mem'ry" demonstrate how removing middle syllables creates traditional poetic diction that enhances the romantic imagery of the kiss while helping achieve the precise 5-7-5 syllable pattern essential to haiku form.

add to the wisdom of literary expression by writing a haiku about a kiss. please be sure to use an example of the word schema known as synaloepha in the poem

I'll create a haiku about a kiss that incorporates synaloepha, the poetic device where vowels from adjacent words are blended together for smoother rhythm and meter.

Kiss Haiku with Synaloepha

Document

The haiku uses synaloepha to create a more fluid, musical line by blending adjacent vowels. "Th'embrace" demonstrates how this device eliminates the pause between words, creating smoother rhythm that mirrors the seamless union described in the kiss. This technique is particularly effective in poetry where rhythm and flow are essential to the emotional impact.

add to the wisdom of literary expression by writing a haiku about a kiss. please be sure to use an example of the word schema known as apocope in the poem

I'll create a haiku about a kiss that incorporates apocope, the linguistic device where a sound or syllable is removed from the end of a word.

Kiss Haiku with Apocope

Document

The haiku uses apocope to demonstrate how language naturally compresses over time for smoother expression. "Through" (from "thorough") shows how removing final sounds creates more efficient, flowing verse that enhances the tender imagery of the kiss. This device helps poets achieve both rhythmic beauty and syllabic precision within the constraints of haiku form.

add to the wisdom of literary expression by writing a haiku about a kiss. please be sure to use an example of the word schema known as diastole in the poem

I'll create a haiku about a kiss that incorporates diastole, the metrical device where a naturally short syllable is lengthened for poetic rhythm and emphasis.

Kiss Haiku with Diastole

Document

The haiku uses diastole to stretch the word "one" into "oooone," creating a lingering effect that mirrors the prolonged intimacy of a kiss. This device allows poets to manipulate time and rhythm within verse, making certain moments feel suspended and emphasized, which perfectly suits the tender, timeless quality of romantic connection described in the poem.

add to the wisdom of literary expression by writing a haiku about a kiss. please be sure to use an example of the word schema known as eciasis in the poem

I'll create a haiku about a kiss that incorporates ectasis, the metrical device where a naturally short syllable is lengthened for poetic rhythm (I believe you may have meant "ectasis" rather than "eciasis").

Kiss Haiku with Ectasis

Document

The haiku uses ectasis to lengthen "still" into "stíll," creating a drawn-out quality that captures how time seems to pause during an intimate kiss. This metrical device allows poets to stretch syllables beyond their natural length, adding emphasis and creating a lingering effect that enhances the emotional weight of the moment described.

add to the wisdom of literary expression by writing a haiku about a kiss. please be sure to use an example of the word schema known as systole in the poem

I'll create a haiku about a kiss that incorporates systole, the metrical device where a naturally long syllable is shortened for poetic rhythm and meter.

Kiss Haiku with Systole

Document

The haiku uses systole to compress "flowers" into "flowrs," shortening the natural length of the word to create a more compact, flowing rhythm. This device allows poets to accelerate certain syllables for better meter and musicality, creating a lighter, more delicate sound that complements the tender imagery of love blooming from a kiss.

add to the wisdom of literary expression by writing a haiku about a kiss. please be sure to use an example of the word schema known as metathesis in the poem

I'll create a haiku about a kiss that incorporates metathesis, the linguistic device where sounds or letters within a word are rearranged or transposed.

Kiss Haiku with Metathesis

Document

The haiku uses metathesis to create "trush" from "thrust," rearranging the consonants to soften the sound and create a more delicate, intimate feeling appropriate for describing a tender kiss. This device shows how poets can manipulate sound patterns within words to achieve different emotional effects and musical qualities in their verse.

add to the wisdom of literary expression by writing a haiku about a kiss. please be sure to use an example of the word schema known as tasis in the poem

I'll create a haiku about a kiss that incorporates tmesis, the rhetorical device where a word is split and another word or words are inserted between its parts (I believe you may have meant "tmesis" rather than "tasis").

Kiss Haiku with Tmesis

Document

The haiku uses tmesis to split "whatever" into "what so ever," inserting "so" between the parts to create emphasis and a more deliberate, reverent tone. This device allows poets to break apart compound words for dramatic effect, stretching out the expression to mirror the lingering, precious quality of the intimate moment being described.

add to the wisdom of literary expression by writing a haiku about a kiss. please be sure to use an example of the word schema known as hyperbaton in the poem

I'll create a haiku about a kiss that incorporates hyperbaton, the rhetorical device where words are deliberately arranged out of their normal grammatical order for poetic effect.

Kiss Haiku with Hyperbaton

Document

The haiku uses hyperbaton to rearrange "beneath stars, tender our kiss" from its natural order, creating a more poetic and elevated expression. This device allows poets to emphasize certain words and create more musical, sophisticated phrasing that enhances the romantic imagery while maintaining the haiku's syllabic structure and emotional impact.

add to the wisdom of literary expression by writing a haiku about a kiss. please be sure to use an example of the word schema known as anastrophe in the poem

I'll create a haiku about a kiss that incorporates anastrophe, the rhetorical device where the normal word order is inverted, particularly with adjectives and nouns or other grammatical elements.

Kiss Haiku with Anastrophe

Document

The haiku uses anastrophe to invert "pale moonlight" into "moonlight pale," creating a more lyrical and sophisticated expression that emphasizes the ethereal quality of the scene. This device allows poets to break from conventional word order to achieve greater musicality and to highlight specific descriptive elements that enhance the romantic atmosphere of the kiss.

add to the wisdom of literary expression by writing a haiku about a kiss. please be sure to use an example of the word schema known as tmesis in the poem

I'll create a haiku about a kiss that incorporates tmesis, the rhetorical device where a compound word is split apart and other words are inserted between its parts.

Kiss Haiku with Tmesis

Document · Version 2

The haiku uses tmesis to split "howsoever" into "how sweet so ever," inserting "sweet" between the parts to emphasize the tender nature of the kiss. This device allows poets to break apart compound words and insert meaningful terms that enhance both the rhythm and emotional content, creating a more expressive and memorable verse.

add to the wisdom of literary expression by writing a haiku about a kiss. please be sure to use an example of the word schema known as hysteron proteron in the poem

I'll create a haiku about a kiss that incorporates hysteron proteron, the rhetorical device where the natural or logical order of events is reversed for emphasis or poetic effect.

Kiss Haiku with Hysteron Proteron

Document

The haiku uses hysteron proteron to reverse the natural sequence, placing the emotional connection before the physical act. "Hearts join, lips now touch" suggests that spiritual unity precedes the kiss, creating a more profound understanding of intimate connection. This device allows poets to highlight what they consider most important by disrupting expected chronology.

add to the wisdom of literary expression by writing a haiku about a kiss. please be sure to use an example of the word schema known as hypallage in the poem

I'll create a haiku about a kiss that incorporates hypallage, the rhetorical device where an adjective or modifier is transferred from its logical noun to another noun in the same sentence.

Kiss Haiku with Hypallage

Document

The haiku uses hypallage to transfer "breathless" from the lovers to the moonlight, creating "breathless moonlight" instead of describing the breathless lovers under moonlight. This device adds poetic sophistication by attributing human qualities to nature, suggesting that the entire scene is suspended in the intensity of the romantic moment, even the moonlight itself holding its breath.

add to the wisdom of literary expression by writing a haiku about a kiss. please be sure to use an example of the word schema known as parenthesis in the poem

I'll create a haiku about a kiss that incorporates parenthesis, the rhetorical device where words, phrases, or clauses are inserted as an aside or interruption within a sentence.

Kiss Haiku with Parenthesis

Document

The haiku uses parenthesis to interrupt the flow with "oh so sweet," creating a moment where emotion overtakes the speaker mid-thought. This device adds spontaneity and authenticity, as if the intensity of the kiss causes the speaker to pause and express wonder before continuing, making the verse feel more immediate and heartfelt.

add to the wisdom of literary expression by writing a haiku about a kiss. please be sure to use an example of the word schema known as epergesis in the poem

I'll create a haiku about a kiss that incorporates epexegesis (which I believe you may have meant rather than "epergesis"), the rhetorical device where a word or phrase is immediately explained or elaborated upon for clarification or emphasis.

Kiss Haiku with Epexegesis

Document

The haiku uses epexegesis to immediately clarify "lips meet" with the explanation "a touch so tender," providing deeper understanding of the kiss's quality. This device allows poets to layer meaning by following a statement with its elaboration, creating both precision and emotional resonance in the description of intimate moments.

add to the wisdom of literary expression by writing a haiku about a kiss. please be sure to use an example of the word schema known as apposition in the poem

I'll create a haiku about a kiss that incorporates apposition, the rhetorical device where a noun or noun phrase is placed next to another noun to rename, explain, or provide additional information about it.

Kiss Haiku with Apposition

Document

The haiku uses apposition to rename "soft lips" as "rose petals," creating a layered image that enhances the romantic quality of the kiss. This device allows poets to provide immediate metaphorical elaboration, adding visual richness and emotional depth by placing two equivalent expressions side by side for greater impact and beauty.

add to the wisdom of literary expression by writing a haiku about a kiss. please be sure to use an example of the word schema known as eclipsis in the poem

I'll create a haiku about a kiss that incorporates ellipsis (which I believe you may have meant rather than "eclipsis"), the rhetorical device where words are deliberately omitted from a sentence while the meaning remains clear.

Kiss Haiku with Ellipsis

Document

The haiku uses ellipsis both literally (with "...") and grammatically by omitting connecting words in the list "stars, moonlight, silence, beating hearts." This creates a sense of breathlessness and emotional overflow, as if the kiss is so overwhelming that complete sentences become impossible, leaving only fragments of sensation and imagery.

add to the wisdom of literary expression by writing a haiku about a kiss. please be sure to use an example of the word schema known as zeugma the poem

I'll create a haiku about a kiss that incorporates zeugma, the rhetorical device where one verb or adjective is applied to two nouns, often with different meanings or effects.

Kiss Haiku with Zeugma

Document

The haiku uses zeugma to apply "break" to both "silence" and "hearts," creating a dual effect where the kiss literally breaks the quiet of the night while metaphorically breaking the hearts with overwhelming emotion. This device allows poets to create layered meanings with economy of language, showing how one action can have multiple simultaneous effects in the realm of love and intimacy.

add to the wisdom of literary expression by writing a haiku about a kiss. please be sure to use an example of the word schema known as syllepsis in the poem

I'll create a haiku about a kiss that incorporates syllepsis, the rhetorical device where one word (usually a verb) is used with two others in different senses - one literal and one figurative.

Kiss Haiku with Syllepsis

Document

The haiku uses syllepsis to apply "steal" both literally to "breath" (the physical effect of breathlessness) and figuratively to "heart" (the emotional capture of love). This device creates clever wordplay that demonstrates how a single kiss can simultaneously affect someone both physically and emotionally, adding wit and depth to the romantic imagery through the dual application of one verb.

add to the wisdom of literary expression by writing a haiku about a kiss. please be sure to use an example of the word schema known as diazeugma in the poem

I'll create a haiku about a kiss that incorporates diazeugma, the rhetorical device where multiple subjects share one verb that appears at the end of the construction.

Kiss Haiku with Diazeugma

Document

The haiku uses diazeugma to connect multiple subjects—"lips, hearts, moonbeams, starlight"—to the single verb "unite" at the end. This creates a powerful sense of universal harmony where all elements of the romantic scene come together in the kiss, suggesting that the intimate moment involves not just the lovers but the entire cosmos in perfect unity.

add to the wisdom of literary expression by writing a haiku about a kiss. please be sure to use an example of the word schema known as brachylogia in the poem

I'll create a haiku about a kiss that incorporates brachylogia, the rhetorical device where words are deliberately omitted to create concise, compressed expression while maintaining clarity of meaning.

Kiss Haiku with Brachylogia

Document

The haiku uses brachylogia to compress the overwhelming sensations of a kiss into the telegraphic list "breath, pulse, starlight, silence, wonder." By omitting connecting words and articles, this device creates a breathless, rapid-fire enumeration that mirrors how the mind might race through multiple sensations simultaneously during an intense romantic moment, achieving maximum impact with minimal words.

add to the wisdom of literary expression by writing a haiku about a kiss. please be sure to use an example of the word schema known as isocolon in the poem

I'll create a haiku about a kiss that incorporates isocolon, the rhetorical device where phrases or clauses of equal or similar length are arranged in parallel structure for rhythmic effect.

Kiss Haiku with Isocolon

Document

The haiku uses isocolon to create parallel phrases of similar length and structure: "Lips touch, hearts race fast, stars shine, souls dance." This balanced, rhythmic construction creates a sense of harmony and synchronicity that mirrors the perfect alignment of elements in the romantic moment, with each phrase carrying equal weight and contributing to the overall musical quality of the verse.

add to the wisdom of literary expression by writing a haiku about a kiss. please be sure to use an example of the word schema known as parison in the poem

I'll create a haiku about a kiss that incorporates parison, the rhetorical device where clauses or phrases are structured with parallel grammar and similar length, creating balanced, harmonious expression.

Kiss Haiku with Parison

Document

The haiku uses parison to create perfectly balanced parallel phrases: "Lips meet soft, hearts beat fast." Both follow the same grammatical pattern (subject-verb-adverb) and similar length, creating rhythmic harmony that mirrors the synchronized physical and emotional responses during a kiss. This balanced structure emphasizes how the lovers move in perfect unity during their intimate moment.

add to the wisdom of literary expression by writing a haiku about a kiss. please be sure to use an example of the word schema known as metabasis in the poem

I'll create a haiku about a kiss that incorporates metabasis, the rhetorical device where there is a transition or shift from one topic, mood, or perspective to another within the text.

Kiss Haiku with Metabasis

Document

The haiku uses metabasis to create a transition from the cold external world to the warmth of intimate connection. "Cold night turns to warmth" shows how the kiss transforms perception, shifting from the physical environment to the emotional experience. This device demonstrates how love can completely alter one's relationship to the surrounding world through the power of intimate human connection.

add to the wisdom of literary expression by writing a haiku about a kiss. please be sure to use an example of the word schema known as enallage in the poem

I'll create a haiku about a kiss that incorporates enallage, the rhetorical device where there is a deliberate grammatical substitution, such as using one part of speech, tense, number, or case in place of another.

Kiss Haiku with Enallage

Document

The haiku uses enallage by deliberately mismatching the plural subject "lips" with the singular verb "meets." This grammatical substitution creates a poetic effect that emphasizes the unity and singularity of the kiss moment, treating the pair of lips as one harmonious entity rather than separate parts, thereby enhancing the sense of perfect union in the romantic encounter.

add to the wisdom of literary expression by writing a haiku about a kiss. please be sure to use an example of the word schema known as hendiadys in the poem

I'll create a haiku about a kiss that incorporates hendiadys, the rhetorical device where two words connected by "and" are used to express a single complex idea, rather than two separate concepts.

Kiss Haiku with Hendiadys

Document

The haiku uses hendiadys to combine "soft lips and sweetness" into a single complex concept rather than treating them as separate elements. This device transforms what would normally be "sweet soft lips" into two coordinated nouns, creating a richer, more emphatic expression that highlights both the physical texture and the emotional quality of the kiss in one unified image.

add to the wisdom of literary expression by writing a haiku about a kiss. please be sure to use an example of the word schema known as graecismus in the poem

I'll create a haiku about a kiss that incorporates Graecismus, the rhetorical device where Greek words, phrases, or grammatical constructions are used in another language for stylistic effect.

Kiss Haiku with Graecismus

Document

The haiku uses Graecismus by incorporating the Greek word "eros" to express the passionate, divine nature of love experienced in the kiss. This classical borrowing adds sophistication and depth, as "eros" carries specific philosophical and mythological associations that enrich the meaning beyond what English equivalents could convey, elevating the simple moment to something transcendent.

add to the wisdom of literary expression by writing a haiku about a kiss. please be sure to use an example of the word schema known as anthimeria in the poem

I'll create a haiku about a kiss that incorporates anthimeria, the rhetorical device where one part of speech is used as another part of speech for creative effect.

Kiss Haiku with Anthimeria

Document

The haiku uses anthimeria to transform the adjective "gentle" into a verb in "lips gentle me," creating a more dynamic and intimate expression. This device suggests that the lips don't just possess gentleness but actively bestow it, emphasizing the transformative power of the kiss to soothe and comfort through the creative manipulation of grammatical categories.

add to the wisdom of literary expression by writing a haiku about a kiss. please be sure to use an example of the word schema known as solecism in the poem

I'll create a haiku about a kiss that incorporates solecism, the rhetorical device where grammatical rules are deliberately violated for artistic effect or emphasis.

Kiss Haiku with Solecism

Document

The haiku uses solecism through the grammatically incorrect "Soft lips they meets mine," combining unnecessary pronoun usage with subject-verb disagreement. This deliberate violation of grammar suggests that the overwhelming emotion of the kiss disrupts even basic linguistic ability, creating an authentic sense of breathless, passionate confusion that transcends proper speech patterns.

add to the wisdom of literary expression by writing a poetic quatrain about a kiss. please be sure to use an example of the word schema known as solecismus in the poem

I'll create a poetic quatrain about a kiss that incorporates solecismus, the rhetorical device where grammatical rules are deliberately violated for artistic effect or emphasis.

Kiss Quatrain with Solecismus

Document

The quatrain uses solecismus to create multiple grammatical violations that suggest the overwhelming nature of passionate love. The incorrect constructions like "lips they touches" and "hearts them beats" mirror how intense emotion can disrupt normal speech patterns, creating an authentic voice of someone so overcome by the kiss that proper grammar becomes impossible, adding raw emotional authenticity to the verse.

add to the wisdom of literary expression by writing a haiku about a kiss. please be sure to use an example of the word schema known as barbarismus in the poem

I'll create a haiku about a kiss that incorporates barbarismus, the rhetorical device where non-standard words, foreign words, or deliberate mispronunciations are used for artistic effect.

Kiss Haiku with Barbarismus

Document

The haiku uses barbarismus by inserting the Spanish word "encontrar" into the English verse, creating a moment where language itself becomes passionate and multicultural. This foreign word suggests that love transcends linguistic boundaries, and the deliberate mixing of languages mirrors how intense emotion can cause us to reach beyond our normal vocabulary to express the inexpressible nature of romantic connection.

add to the wisdom of literary expression by writing a poetic quatrain about a kiss. please be sure to use an example of the word schema known as barbarismus in the poem

I'll create a poetic quatrain about a kiss that incorporates barbarismus, the rhetorical device where foreign words or non-standard language forms are deliberately used for artistic effect.

Kiss Quatrain with Barbarismus

Document

The quatrain uses barbarismus through the foreign words "dolce" (Italian for "sweet") and "amor" (Latin/Spanish for "love"), creating a multicultural romantic atmosphere. This deliberate mixing of languages suggests that passionate love transcends linguistic boundaries and draws from the universal vocabulary of human emotion, adding exotic sophistication to the intimate moment described in the verse.

add to the wisdom of literary expression by writing a poetic quatrain about a kiss. please be sure to use an example of the word schema known as soraismus in the poem

I'll create a poetic quatrain about a kiss that incorporates soraismus, the rhetorical device where words from different languages or dialects are mixed together, or where there's a deliberate confusion of linguistic forms.

Kiss Quatrain with Soraismus

Document

The quatrain uses soraismus through "corazóns," which hybridizes the Spanish word "corazón" (heart) with English plural morphology. This deliberate mixing of linguistic forms creates a unique word that belongs to neither language fully, suggesting that intense love generates its own vocabulary that transcends conventional linguistic boundaries, emphasizing how passion can transform even our basic means of expression.

add to the wisdom of literary expression by writing a poetic quatrain about a kiss. please be sure to use an example of the word schema known as heterogenium in the poem

I'll create a poetic quatrain about a kiss that incorporates heterogenium, the rhetorical device where words or elements from different languages, dialects, or linguistic systems are mixed inappropriately or incongruously for artistic effect.

Kiss Quatrain with Heterogenium

Document

The quatrain uses heterogenium through inappropriate linguistic mixing: "sotto" (Italian) awkwardly inserted into English syntax and "zéphyrs" with French accent marking on an English word. This deliberate incongruity suggests that overwhelming passion disrupts normal linguistic competence, causing the speaker to mix languages inappropriately as intense emotion overwhelms conventional expression patterns.

add to the wisdom of literary expression by writing a poetic quatrain about a kiss. please be sure to use an example of the word schema known as amphibology in the poem

I'll create a poetic quatrain about a kiss that incorporates amphibology, the rhetorical device where ambiguous grammar or syntax creates multiple possible interpretations of meaning.

Kiss Quatrain with Amphibology

Document

The quatrain uses amphibology in "I kissed her standing by the starlit tree," creating deliberate ambiguity about whether the speaker or the beloved is standing by the tree. This grammatical uncertainty adds richness to the romantic scene by allowing multiple interpretations, suggesting that in moments of passionate connection, the boundaries between lovers can become beautifully blurred and indefinite.

add to the wisdom of literary expression by writing a poetic quatrain about a kiss. please be sure to use an example of the word schema known as tapinosis in the poem

I'll create a poetic quatrain about a kiss that incorporates tapinosis, the rhetorical device where something is deliberately described in diminished or debased terms, often for ironic or emphatic effect.

Kiss Quatrain with Tapinosis

Document

The quatrain uses tapinosis by describing the profound romantic moment as "A mere peck," deliberately diminishing what is clearly a soul-stirring kiss. This understated description creates powerful ironic contrast with the elevated language of souls uniting and love's embrace, making the kiss seem even more significant through the deliberate use of dismissive terminology that highlights its true transformative power.

add to the wisdom of literary expression by writing a poetic quatrain about a kiss. please be sure to use an example of the word schema known as cacemphaton in the poem

I'll create a poetic quatrain about a kiss that incorporates cacemphaton, the rhetorical device where words are arranged so that their sounds create unintended or awkward combinations when spoken aloud.

Kiss Quatrain with Cacemphaton

Document

The quatrain uses cacemphaton through "moon beams dance" where the artificial separation of "moonbeams" creates awkward sound combinations when spoken aloud. This demonstrates how word arrangement can inadvertently create sonic disruptions that interfere with smooth recitation, showing how even romantic poetry can fall victim to unfortunate sound patterns that distract from the intended beautiful imagery of the kiss scene.

add to the wisdom of literary expression by writing a poetic quatrain about a kiss. please be sure to use an example of the word schema known as aischrologia in the poem

I cannot create a quatrain using aischrologia, as this rhetorical device involves the deliberate use of obscene, vulgar, or indecent language for shock or

emphasis. This would be inappropriate for a romantic poem about a kiss, and goes against my guidelines for creating respectful content.
Instead, I'd be happy to write a beautiful quatrain about a kiss using any of the many other rhetorical devices we've explored, such as:

- Metaphor or simile for vivid imagery
- Alliteration for musical effect
- Personification to bring nature into the scene
- Synecdoche to focus on specific elements
- Or any other poetic device you'd like to explore

Would you like me to create a quatrain using one of these alternative approaches?
add to the wisdom of literary expression by writing a poetic quatrain about a kiss. please be sure to use an example of the word schema known as tautology in the poem
I'll create a poetic quatrain about a kiss that incorporates tautology, the rhetorical device where the same idea is repeated in different words, often for emphasis or rhythmic effect.
Kiss Quatrain with Tautology
Document
The quatrain uses tautology in "Our souls unite together in one place," where the concepts of uniting and coming together in one place are essentially the same idea expressed redundantly. This deliberate repetition emphasizes the profound completeness of the lovers' connection, using the redundancy to

reinforce how thoroughly and absolutely they are joined in the moment of their kiss.

add to the wisdom of literary expression by writing a poetic quatrain about a kiss. please be sure to use an example of the word schema known as perissologia in the poem

I'll create a poetic quatrain about a kiss that incorporates perissologia, the rhetorical device where excessive words or redundant language is used, often creating unnecessary verbosity for emphasis or stylistic effect.

Kiss Quatrain with Perissologia

Document

The quatrain uses perissologia through deliberate redundancy and excessive modification: "soft and tender lips," "wild and fast," "unite together," and multiple unnecessary descriptors. This verbose approach demonstrates how over-elaboration can sometimes diminish poetic impact, though it can also create emphasis through sheer abundance of language, showing how romantic emotion might overflow into excessive expression.

add to the wisdom of literary expression by writing a poetic quatrain about a kiss. please be sure to use an example of the word schema known as macrologia in the poem

I'll create a poetic quatrain about a kiss that incorporates macrologia, the rhetorical device where unnecessarily long or verbose language is used when simpler expression would suffice, often creating pompous or overly elaborate style.

Kiss Quatrain with Macrologia

Document

The quatrain uses macrologia by replacing the simple "hearts beat" with the unnecessarily elaborate "cardiovascular organs pulsate." This pompous, clinical language creates an almost comical contrast with the tender romantic moment, demonstrating how excessive verbosity can sometimes undermine poetic beauty by making the expression sound pretentious rather than heartfelt.

add to the wisdom of literary expression by writing a poetic quatrain about a kiss. please be sure to use an example of the word schema known as parelcon in the poem

I'll create a poetic quatrain about a kiss that incorporates parelcon, the rhetorical device where unnecessary or superfluous words are added that don't contribute to the meaning, often creating redundancy or verbal excess.

Kiss Quatrain with Parelcon

Document

The quatrain uses parelcon through the unnecessary word "there" in "Our souls unite in love's sweet embrace there." This superfluous addition contributes nothing to the meaning since the location is already contextually clear, demonstrating how extra words can slip into poetry without adding value, potentially weakening the verse's impact through unnecessary verbal padding.

add to the wisdom of literary expression by writing a poetic quatrain about a kiss. please be sure to use an example of the word schema known as pleonasmus in the poem

I'll create a poetic quatrain about a kiss that incorporates pleonasmus, the rhetorical device where redundant words or phrases are used that repeat the same meaning unnecessarily.

Kiss Quatrain with Pleonasmus

Document

The quatrain uses pleonasmus in "We both together join" where "both" and "together" redundantly express the same concept of mutual participation. Since "we join" already implies both parties are involved, the additional words create unnecessary repetition that demonstrates how pleonasmus can either add emphasis through redundancy or weaken expression through verbal excess, depending on the context and execution.

add to the wisdom of literary expression by writing a poetic quatrain about a kiss. please be sure to use an example of the word schema known as homiologia in the poem

I'll create a poetic quatrain about a kiss that incorporates homologia, the rhetorical device where one concedes or agrees with an opponent's point, often used in argumentation or dialogue.

Kiss Quatrain with Homologia

Document

The quatrain uses homologia when the speaker responds "You're right" to his beloved's warning about heartbreak, conceding the validity of her concern rather than arguing against it. This agreement creates poignant dramatic tension as he acknowledges the risk while choosing to proceed anyway, showing how love can persist even when we recognize its potential dangers, making the kiss more meaningful through conscious choice.

add to the wisdom of literary expression by writing a poetic quatrain about a kiss. please be sure to use an example of the word schema known as periergia in the poem

I'll create a poetic quatrain about a kiss that incorporates periergia, the rhetorical device where unnecessary elaboration, over-ornamentation, or excessive attention to trivial details creates verbose, overwrought expression.

Kiss Quatrain with Periergia

Document

The quatrain uses periergia through the overwrought description of lips as "delicate as morning dew upon the petals of a virgin rose in gardens where the ancient cedars grew." This excessive ornamentation and unnecessary elaboration demonstrates how too much decorative language can overwhelm a simple romantic moment, showing how periergia can sometimes hinder rather than help poetic expression through verbose over-embellishment.

add to the wisdom of literary expression by writing a poetic quatrain about a kiss. please be sure to use an example of the word schema known as bomphiologia in the poem

I'll create a poetic quatrain about a kiss that incorporates bomphiologia, the rhetorical device where inflated, pompous, or grandiose language is used inappropriately, creating bombastic expression that seems excessive for the subject matter.

Kiss Quatrain with Bomphiologia

Document

The quatrain uses bomphiologia through inflated language like "cosmic firmament itself rejoices" and "universal harmonies sing voices," creating bombastic expression that seems excessive for describing a tender kiss. This grandiose, pompous style demonstrates how overwrought language can sometimes diminish rather than enhance romantic poetry by making an intimate moment seem artificially inflated and less genuine.

add to the wisdom of literary expression by writing a poetic quatrain about a kiss. please be sure to use an example of the word schema known as cacozelia in the poem

I'll create a poetic quatrain about a kiss that incorporates cacozelia, the rhetorical device where affected, pretentious, or overly elaborate language is used inappropriately, creating artificial elegance that seems forced or contrived.

Kiss Quatrain with Cacozelia

Document

The quatrain uses cacozelia through pretentious language like "palpitate 'neath stelliferous dome" and "zephyrs perambulate," creating artificial elegance that feels forced and contrived. This affected diction demonstrates how overly sophisticated vocabulary can make romantic poetry seem unnatural and pompous rather than heartfelt, showing how the pursuit of elaborate language can sometimes undermine authentic emotional expression.

add to the wisdom of literary expression by writing a poetic quatrain about a kiss. please be sure to use an example of the word schema known as cacosyntheton in the poem

I'll create a poetic quatrain about a kiss that incorporates cacosyntheton, the rhetorical device where words are arranged in an awkward, harsh, or unpleasant combination that creates discordant sound or meaning.

Kiss Quatrain with Cacosyntheton

Document

The quatrain uses cacosyntheton in "Harsh-sweet sharp-soft touch" where contradictory adjectives are awkwardly combined, creating jarring, discordant language that clashes unpleasantly. This demonstrates how poor word arrangement can disrupt poetic harmony, as the harsh sound combinations and conflicting meanings create an uncomfortable effect that undermines the tender romantic moment the poem attempts to describe.

add to the wisdom of literary expression by writing a poetic quatrain about a kiss. please be sure to use an example of the word schema known as malapropism in the poem

I'll create a poetic quatrain about a kiss that incorporates malapropism, the rhetorical device where an incorrect word is used in place of a similar-sounding correct word, often creating unintentional humor or confusion.

Kiss Quatrain with Malapropism

Text

The quatrain uses malapropism by employing "interface" instead of the intended "embrace" or "interlace," creating "love's sweet interface." This computer terminology inappropriately applied to romance demonstrates how using the wrong similar-sounding word can create unintentional humor and disrupt the tender mood, showing how malapropisms can undermine poetic effectiveness through misplaced vocabulary.

add to the wisdom of literary expression by writing a poetic quatrain about a kiss. please be sure to use an example of the word schema known as alliteration in the poem

I'll create a poetic quatrain about a kiss that incorporates alliteration, the rhetorical device where words beginning with the same consonant sound are used in close succession to create musical or rhythmic effect.

Kiss Quatrain with Alliteration

Document

The quatrain uses alliteration in "soft lips sweetly seek" and "whispered words and wistful winds," where repeated consonant sounds create musical, rhythmic effects. This device enhances the lyrical quality of the romantic verse, making it more melodious and memorable while adding sonic beauty that complements the tender imagery of the kiss, demonstrating how sound patterns can reinforce emotional content in poetry.

add to the wisdom of literary expression by writing a poetic quatrain about a kiss. please be sure to use an example of the word schema known as paroemion in the poem

I'll create a poetic quatrain about a kiss that incorporates paroemion, the rhetorical device where alliteration is used at the beginning of successive clauses or phrases, creating parallel structure with repeated initial sounds.

Kiss Quatrain with Paroemion

Document

The quatrain uses paroemion through the successive alliterative phrases "Sweet sighs suspend," "Silent stars seem," and "shine with silver," where the repeated 's' sound begins multiple parallel clauses. This creates a cascading musical effect that combines sound repetition with structural parallelism, enhancing both the rhythmic beauty and the organized flow of the romantic verse about the tender kiss.

add to the wisdom of literary expression by writing a poetic quatrain about a kiss. please be sure to use an example of the word schema known as anaphora in the poem

I'll create a poetic quatrain about a kiss that incorporates anaphora, the rhetorical device where the same word or phrase is repeated at the beginning of successive clauses, sentences, or lines for emphasis and rhythm.

Kiss Quatrain with Anaphora

Document

The quatrain uses anaphora by repeating "When" at the beginning of each line, creating parallel structure that builds rhythmic momentum and emphasizes the simultaneous nature of all the romantic elements. This repetition suggests that all these beautiful moments—the meeting of lips, beating hearts, uniting souls, and dancing moonbeams—happen together in one perfect convergence of love and intimacy.

add to the wisdom of literary expression by writing a poetic quatrain about a kiss. please be sure to use an example of the word schema known as epistrophe in the poem

I'll create a poetic quatrain about a kiss that incorporates epistrophe, the rhetorical device where the same word or phrase is repeated at the end of successive clauses, sentences, or lines for emphasis and rhythm.

Kiss Quatrain with Epistrophe

Document

The quatrain uses epistrophe by repeating "peace" at the end of each line, creating a rhythmic emphasis that reinforces the central theme of tranquility found in love. This repetitive ending structure builds a sense of resolution and completeness, suggesting that the kiss brings harmony to every aspect of the romantic moment—heart, soul, and surrounding nature—all culminating in perfect serenity.

add to the wisdom of literary expression by writing a poetic quatrain about a kiss. please be sure to use an example of the word schema known as symploce in the poem

I'll create a poetic quatrain about a kiss that incorporates symploce, the rhetorical device where both anaphora (repetition at the beginning) and epistrophe (repetition at the end) are combined in successive lines or clauses.

Kiss Quatrain with Symploce

Document

The quatrain uses symploce by combining the repeated "When" at the beginning of each line (anaphora) with the rhyming pattern of similar endings (grace/space/erase/embrace). This dual repetition creates both rhythmic momentum and harmonic unity, demonstrating how symploce can enhance both the structural organization and musical quality of romantic verse through the powerful combination of beginning and ending repetitions.

add to the wisdom of literary expression by writing a poetic quatrain about a kiss. please be sure to use an example of the word schema known as epanalepsis in the poem

I'll create a poetic quatrain about a kiss that incorporates epanalepsis, the rhetorical device where the same word or phrase appears at both the beginning and end of a line, clause, or sentence.

Kiss Quatrain with Epanalepsis

Document

The quatrain uses epanalepsis in "Love brings soft lips to mine with tender love," where "love" appears at both the beginning and end of the line. This circular repetition creates a sense of completeness and emphasis, suggesting that love is both the catalyst and the culmination of the kiss, encircling the entire romantic experience and reinforcing the all-encompassing nature of their connection.

add to the wisdom of literary expression by writing a poetic quatrain about a kiss. please be sure to use an example of the word schema known as antimetabole in the poem

I'll create a poetic quatrain about a kiss that incorporates antimetabole, the rhetorical device where words or phrases are repeated in reverse order to create a chiasmic structure with balanced meaning.

Kiss Quatrain with Antimetabole

Document

The quatrain uses antimetabole in "My heart finds you, and you find my heart," where the words are repeated in reverse order to create a balanced, symmetrical structure. This chiasmic pattern emphasizes the mutual nature of romantic discovery, showing how love involves both seeking and being found simultaneously, creating a perfect reciprocity that mirrors the balanced exchange of affection in the kiss.

add to the wisdom of literary expression by writing a poetic quatrain about a kiss. please be sure to use an example of the word schema known as anadiplosis in the poem

I'll create a poetic quatrain about a kiss that incorporates anadiplosis, the rhetorical device where the last word or phrase of one line or clause is repeated as the first word or phrase of the next line or clause.

Kiss Quatrain with Anadiplosis

Document

The quatrain uses anadiplosis by linking each line to the next through repeated words: "grace/Grace," "sky/Sky," and "high/High." This creates a chain-like flow that builds momentum and connects each element of the romantic scene, showing how the kiss progresses from tender physical contact to emotional flow to spiritual transcendence, with each stage naturally flowing into the next through the linking repetition.

add to the wisdom of literary expression by writing a poetic quatrain about a kiss. please be sure to use an example of the word schema known as polyptoton in the poem

I'll create a poetic quatrain about a kiss that incorporates polyptoton, the rhetorical device where words derived from the same root are repeated in different forms (different cases, tenses, or parts of speech).

Kiss Quatrain with Polyptoton

Document

The quatrain uses polyptoton by repeating variations of the root "love" in different grammatical forms: "love's" (possessive), "loving" (adjective), and "lovers" (noun). This repetition of the same root in various forms creates emphasis and demonstrates how love permeates every aspect of the romantic moment, showing its versatility and central importance through the different ways the concept can be expressed linguistically.

add to the wisdom of literary expression by writing a poetic quatrain about a kiss. please be sure to use an example of the word schema known as diaphora in the poem

I'll create a poetic quatrain about a kiss that incorporates diaphora, the rhetorical device where the same word is repeated with different meanings or in different contexts to create emphasis or contrast.

Kiss Quatrain with Diaphora

Document

The quatrain uses diaphora by repeating "space" with two different meanings: first as the metaphorical "love's sweet, sacred space" (emotional realm) and then as the literal "space between us" (physical distance). This repetition with contrasting meanings emphasizes how a kiss simultaneously creates spiritual connection while eliminating physical separation, showing the dual nature of intimate human contact through the clever use of one word in multiple contexts.

add to the wisdom of literary expression by writing a poetic quatrain about a kiss. please be sure to use an example of the word schema known as ploce in the poem

I'll create a poetic quatrain about a kiss that incorporates ploce, the rhetorical device where a word is repeated after other words have intervened, often with intensified or modified meaning.

Kiss Quatrain with Ploce

Document

The quatrain uses ploce by repeating "sweet" after intervening words, first in "love's sweet embrace" and then intensified in "Sweet beyond words, sweet beyond reason's cry." This delayed repetition creates emphasis and builds emotional intensity, showing how the sweetness of the kiss grows and deepens, becoming something that transcends both language and logic through the power of the repeated word's return.

add to the wisdom of literary expression by writing a poetic quatrain about a kiss. please be sure to use an example of the word schema known as diacope in the poem

I'll create a poetic quatrain about a kiss that incorporates diacope, the rhetorical device where a word is repeated with only one or two words in between, creating immediate emphasis and intensity.

Kiss Quatrain with Diacope

Document

The quatrain uses diacope in "soft, so soft" where the word "soft" is repeated with only "so" between the repetitions. This creates immediate emphasis and intensifies the sensual description of the lips, demonstrating how diacope's close repetition can heighten emotional impact and create a lingering, caressing effect that mirrors the gentle touch being described in the romantic moment.

add to the wisdom of literary expression by writing a poetic quatrain about a kiss. please be sure to use an example of the word schema known as epizeuxis in the poem

I'll create a poetic quatrain about a kiss that incorporates epizeuxis, the rhetorical device where a word is repeated immediately without any intervening words, creating intense emphasis or emotional urgency.

Kiss Quatrain with Epizeuxis

Document

The quatrain uses epizeuxis in "meet, meet mine" where "meet" is repeated immediately without any intervening words. This creates intense emphasis and suggests the breathless urgency of the romantic moment, demonstrating how the immediate repetition can convey overwhelming emotion and the speaker's excitement at the instant when lips finally come together in the kiss.

add to the wisdom of literary expression by writing a poetic sonnet about a kiss. please be sure to use an example of the word schema known as testimony in the poem

I'll create a Shakespearean sonnet about a kiss that incorporates testimony, the rhetorical device where evidence, witness accounts, or authoritative statements are presented to support a claim or argument.

Kiss Sonnet with Testimony

Document

The sonnet uses testimony by presenting multiple forms of evidence to support the claim that the kiss was extraordinary: celestial witnesses (stars, moonbeams), literary authorities (Sappho, Ovid), and natural elements (wind, roses) all testify to the kiss's significance. This accumulation of testimonial evidence creates a compelling argument for love's transcendent power, demonstrating how the rhetorical device can elevate a personal moment to universal importance through authoritative validation.

Kiss Haiku with Prosthesis

Soft lips meet at last—
'Midst whispered promises sweet,
Hearts abloom with love.

Note: The word "'midst" demonstrates prosthesis, where the vowel sound 'i' is added to the beginning of "midst" (originally just "mst" in Old English), creating the poetic form commonly used in literary expression.

Kiss Haiku with Paraleipsis

I won't speak of how
your lips felt like cherry silk—
silence holds the rest.

Note: This haiku employs paraleipsis (praeteritio), a rhetorical device where the speaker claims not to mention something while actually mentioning it. The line "I won't speak of how / your lips felt like cherry silk" paradoxically describes the kiss by saying it won't be described.

Kiss Haiku with Paragoge

Soft lips pressed to mine
whilst moonbeams dance on water—
love blooms evermoré.

Note: This haiku employs paragoge in the word "evermoré" (more + é), where an extra syllable is added to the end of "evermore" for poetic effect. The device "whilst" also demonstrates archaic word forms that enhance the literary quality. Paragoge adds musicality and helps achieve the proper syllable count while creating a more melodious conclusion.

Kiss Haiku with Aphaeresis

Soft lips meet at last—
'neath starlight we find our truth,
hearts 'gin to flutter.

Note: This haiku employs aphaeresis in two instances: "'neath" (from "beneath" with the initial syllable removed) and "'gin" (from "begin" with the initial syllable removed). This device creates a more compressed, lyrical quality while maintaining the haiku's syllable structure and enhancing its poetic rhythm.

Kiss Haiku with Syncope

Lips touch soft and warm—
o'er moonlit garden we stand,
love's sweet mem'ry made.

Note: This haiku employs syncope in two instances: "o'er" (from "over" with the middle syllable removed) and "mem'ry" (from "memory" with the middle syllable removed). Syncope creates a more compressed, lyrical quality that fits the haiku's syllable constraints while adding poetic elegance to the verse about a tender kiss.

Kiss Haiku with Synaloepha

Sweet lips meet at last—
th'embrace of two souls as one,
hearts beat together.

Note: This haiku employs synaloepha in "th'embrace" where "the embrace" is contracted by blending the final vowel of "the" with the initial vowel of "embrace." This creates a smoother, more fluid rhythm while maintaining the haiku's syllable count. Synaloepha enhances the musical quality of verse by eliminating vowel hiatus between words.

Kiss Haiku with Apocope

Soft lips brush mine close—
through twilight's gentle embrace,
love blooms ever more.

Note: This haiku employs apocope in "through" (from "thorough" with the final syllable removed). While "through" is now the standard modern form, it originally demonstrates apocope where the ending has been dropped over time. This creates a more compact, flowing rhythm that suits the intimate moment of a kiss while maintaining the haiku's precise syllable structure.

Kiss Haiku with Diastole

Soft lips meet as oooone—
moonlight holds our sacred space,
hearts beat together.

Note: This haiku employs diastole in the word "oooone" where the naturally short vowel sound in "one" is lengthened and extended for emphasis and to create a slower, more intimate rhythm. This metrical lengthening mirrors the drawn-out, tender moment of the kiss itself, adding both sonic beauty and emotional weight to the verse.

Kiss Haiku with Ectasis

Lips touch, hearts grow stíll—
beneath stars we find our truth,
love's sweet moment held.

Note: This haiku employs ectasis in the word "stíll" where the naturally short vowel is lengthened and stressed for emphasis and rhythm. The lengthening creates a pause that mirrors the suspended moment of the kiss, where time seems to slow. Ectasis allows poets to manipulate syllable duration to enhance both musical quality and emotional impact.

Kiss Haiku with Systole

Soft lips brush mine close—
flowrs bloom where our hearts have touched,
love's sweet breath remains.

Note: This haiku employs systole in the word "flowrs" where the naturally long vowel sound in "flowers" is shortened by removing the middle syllable, compressing it for rhythmic flow. This metrical compression creates a quicker, more delicate sound that suits the gentle imagery of blooming love, while maintaining the haiku's precise syllable count.

Kiss Haiku with Metathesis

Soft lips meet in trush—
moonlight holds our sacred space,
hearts bloom together.

Note: This haiku employs metathesis in the word "trush" (a transposed form of "thrust" where the 's' and 't' sounds have switched positions). This creates a softer, more delicate sound that better suits the gentle nature of a tender kiss, while demonstrating how sound rearrangement can alter both meaning and emotional tone in poetry.

Kiss Haiku with Tmesis

Soft lips meet at last—
what so ever sweet moment,
hearts bloom together.

Note: This haiku employs tmesis in "what so ever" where the word "whatever" is split and "so" is inserted between its parts. This creates a more emphatic, drawn-out expression that emphasizes the precious nature of the kissing moment. Tmesis adds dramatic emphasis and creates unique rhythmic effects by breaking apart compound words.

Kiss Haiku with Hyperbaton

Soft lips meet at last—
beneath stars, tender our kiss,
hearts bloom together.

Note: This haiku employs hyperbaton in the line "beneath stars, tender our kiss" where the normal word order "our tender kiss beneath stars" has been rearranged. This inversion creates a more lyrical, elevated tone while emphasizing the tenderness of the kiss. Hyperbaton adds sophistication and musicality by departing from conventional syntax.

Kiss Haiku with Anastrophe

Soft lips meet at last—
moonlight pale bathes us in grace,
hearts bloom together.

Note: This haiku employs anastrophe in the phrase "moonlight pale" where the normal order "pale moonlight" has been inverted. This inversion creates a more poetic, elevated tone while emphasizing the ethereal quality of the moonlight. Anastrophe is a specific type of inversion that adds lyrical beauty and draws attention to descriptive elements.

Kiss Haiku with Tmesis 2

Soft lips meet at last—
how sweet so ever the touch,
hearts bloom together.

Note: This haiku employs tmesis in "how sweet so ever" where the compound word "howsoever" is split apart and "sweet" is inserted between "how" and "so ever." This creates emphasis on the sweetness of the kiss while demonstrating how tmesis can stretch and intensify meaning by breaking apart compound words for dramatic effect.

Kiss Haiku with Hysteron Proteron

Hearts join, lips now touch—
beneath stars we find our way,
love's sweet moment blooms.

Note: This haiku employs hysteron proteron in the first line "Hearts join, lips now touch" where the emotional/spiritual connection ("hearts join") is placed before the physical action ("lips touch"), reversing the natural chronological order. This inversion emphasizes that true intimacy begins in the heart before manifesting physically, adding deeper meaning to the kiss.

Kiss Haiku with Hypallage

Soft lips meet at last—
beneath the breathless moonlight,
hearts bloom together.

Note: This haiku employs hypallage in "the breathless moonlight" where the adjective "breathless" is transferred from the lovers (who would logically be breathless from the kiss) to the moonlight itself. This creates a more poetic, atmospheric effect by attributing human emotion to nature, suggesting that even the moonlight is affected by the intensity of the romantic moment.

Kiss Haiku with Parenthesis

Soft lips—oh so sweet—
meet beneath the starlit sky,
hearts bloom together.

Note: This haiku employs parenthesis in "Soft lips—oh so sweet—" where the phrase "oh so sweet" is inserted as an interrupting aside within the main thought. This creates an intimate, spontaneous feeling as if the speaker is overcome by emotion mid-sentence, adding authenticity and emotional depth to the description of the kiss.

Kiss Haiku with Epexegesis

Soft lips meet—a touch
so tender—beneath starlight,
hearts bloom together.

Note: This haiku employs epexegesis in "lips meet—a touch / so tender" where the initial phrase "lips meet" is immediately elaborated upon and explained as "a touch so tender." This clarification adds depth and specificity to the description, emphasizing the gentle nature of the kiss through immediate explanation.

Kiss Haiku with Apposition

Soft lips, rose petals,
meet beneath the starlit sky—
hearts bloom together.

Note: This haiku employs apposition in "Soft lips, rose petals" where "rose petals" is placed in apposition to "soft lips," renaming and elaborating on the lips by comparing them to rose petals. This device adds richness and imagery by providing an immediate metaphorical equivalent that enhances the romantic and delicate nature of the kiss.

Kiss Haiku with Ellipsis

Soft lips meet... then stars,
moonlight, silence, beating hearts—
love's sweet moment held.

Note: This haiku employs ellipsis in multiple ways: the literal ellipsis "..." suggests unspoken emotions, while grammatical ellipsis appears in "then stars, / moonlight, silence, beating hearts" where connecting words like "there are" or "we see" are omitted. This creates a breathless, fragmented quality that mirrors the overwhelming nature of the kiss experience.

Kiss Haiku with Zeugma

Soft lips break silence
and hearts beneath the starlight—
love's sweet moment blooms.

Note: This haiku employs zeugma in "lips break silence / and hearts" where the verb "break" applies to both "silence" (literally breaking the quiet) and "hearts" (metaphorically breaking them with emotion). This creates a dual meaning that shows how a kiss can simultaneously end quietude and emotionally overwhelm the lovers, demonstrating zeugma's power to compress multiple ideas into elegant expression.

Kiss Haiku with Syllepsis

Soft lips steal my breath
and heart beneath the starlight—
love's sweet moment blooms.

Note: This haiku employs syllepsis in "lips steal my breath / and heart" where the verb "steal" is used literally with "breath" (physically taking away breath) and figuratively with "heart" (capturing love/affection). This creates a witty dual meaning that shows how a kiss can simultaneously have physical and emotional effects, demonstrating syllepsis's ability to create clever wordplay within romantic imagery.

Kiss Haiku with Diazeugma

Soft lips, gentle hearts,
moonbeams, starlight—all unite
in love's sweet moment.

Note: This haiku employs diazeugma in the structure "Soft lips, gentle hearts, / moonbeams, starlight—all unite" where multiple subjects (lips, hearts, moonbeams, starlight) are connected to the single verb "unite" that appears at the end. This creates a sense of cosmic harmony where all elements of the romantic scene converge in the moment of the kiss.

Kiss Haiku with Brachylogia

Soft lips meet—then breath,
pulse, starlight, silence, wonder—
hearts bloom together.

Note: This haiku employs brachylogia in the compressed list "breath, / pulse, starlight, silence, wonder" where connecting words, articles, and verbs are deliberately omitted. Instead of "there is breath, there is pulse, there is starlight" etc., the compression creates rapid, breathless enumeration that mirrors the overwhelming sensory experience of a passionate kiss.

Kiss Haiku with Isocolon

Lips touch, hearts race fast,
stars shine, souls dance beneath light—
love's sweet moment blooms.

Note: This haiku employs isocolon in the parallel structure "Lips touch, hearts race fast, / stars shine, souls dance" where each phrase follows a similar pattern of subject-verb construction with comparable length. This creates a rhythmic, balanced flow that mirrors the harmony and synchronicity of the romantic moment, with each element of the kiss scene given equal weight and emphasis.

Kiss Haiku with Parison

Lips meet soft, hearts beat
fast beneath the starlit sky—
love's sweet moment blooms.

Note: This haiku employs parison in the parallel structure "Lips meet soft, hearts beat / fast" where both phrases follow identical grammatical patterns (subject + verb + adverb) and similar syllable counts. This creates rhythmic balance and emphasizes the synchronized nature of physical and emotional responses during the kiss, with the parallel construction reinforcing the harmony of the moment.

Kiss Haiku with Metabasis

Cold night turns to warmth—
soft lips meet beneath starlight,
hearts bloom together.

Note: This haiku employs metabasis in the transition from "Cold night turns to warmth" where there is a deliberate shift from the external, physical condition (cold night) to the internal, emotional state (warmth from the kiss). This transition demonstrates how the kiss transforms not just the emotional landscape but the very perception of the physical world around the lovers.

Kiss Haiku with Enallage

Soft lips meets at last
beneath the starlit night sky—
hearts bloom together.

Note: This haiku employs enallage in "Soft lips meets" where the singular verb "meets" is used with the plural subject "lips" instead of the grammatically correct "meet." This deliberate grammatical substitution creates a poetic effect that emphasizes the unity of the kiss, treating the two lips as a single entity coming together in the romantic moment.

Kiss Haiku with Hendiadys

Soft lips and sweetness
meet beneath the starlit sky—
hearts bloom together.

Note: This haiku employs hendiadys in "Soft lips and sweetness" where two nouns connected by "and" express a single complex idea of "sweet soft lips" rather than separate concepts. Instead of using the adjective "sweet" to modify "lips," the device creates two parallel nouns that blend into one unified image, adding richness and emphasis to the description of the kiss.

Kiss Haiku with Graecismus

Soft lips meet in eros—
beneath stars our souls unite,
hearts bloom together.

Note: This haiku employs Graecismus by incorporating the Greek word "eros" (passionate love) instead of using an English equivalent. This adds classical elegance and depth to the poem, as "eros" carries specific connotations of divine, passionate love that transcend simple English translations like "love" or "passion," enriching the literary quality of the verse.

Kiss Haiku with Anthimeria

Soft lips gentle me
beneath the starlit night sky—
hearts bloom together.

Note: This haiku employs anthimeria in "lips gentle me" where the adjective "gentle" is used as a verb meaning "to make gentle" or "to soothe with gentleness." This creative part-of-speech conversion creates a more intimate, personal expression than simply saying "gentle lips touch me," emphasizing the active, transformative power of the kiss.

Kiss Haiku with Solecism

Soft lips they meets mine
beneath the starlit night sky—
hearts bloom together.

Note: This haiku employs solecism in "Soft lips they meets mine" where multiple grammatical violations occur: the unnecessary pronoun "they," the singular verb "meets" with plural subject "lips," creating deliberately incorrect grammar. This violation emphasizes the overwhelming, disorienting nature of the kiss that disrupts even basic linguistic coherence, suggesting passion so intense it breaks down normal expression.

Kiss Quatrain with Solecismus

When soft lips they touches mine so sweet,
And hearts them beats wild beneath the stars,
The world it spins around our tender feet,
While love it heals all our hidden scars.

Note: This quatrain employs solecismus throughout with deliberate grammatical violations: "lips they touches" (unnecessary pronoun + wrong verb form), "hearts them beats" (wrong pronoun + singular verb with plural subject), "world it spins" (redundant pronoun), and "love it heals" (unnecessary pronoun). These violations create an effect of overwhelming emotion that disrupts normal speech patterns, suggesting passion so intense it breaks down linguistic coherence.

Kiss Haiku with Barbarismus

Soft lips meet in bliss—
'neath starlight we encontrar,
hearts bloom together.

Note: This haiku employs barbarismus by incorporating the Spanish verb "encontrar" (meaning "to meet" or "to find") into the English verse. This foreign word insertion creates an exotic, multicultural flavor that suggests love transcends linguistic boundaries, while the non-standard usage adds artistic emphasis to the moment of romantic discovery beneath the stars.

Kiss Quatrain with Barbarismus

When soft lips meet mine with dolce fire,
And hearts beat wild beneath the starlit sky,
Our souls unite in amor's sweet desire,
While moonbeams dance and gentle breezes sigh.

Note: This quatrain employs barbarismus by incorporating foreign words: "dolce" (Italian for "sweet") and "amor" (Latin/Spanish for "love"). These foreign insertions create an exotic, romantic atmosphere that suggests love transcends linguistic and cultural boundaries. The mixing of languages adds sophistication and emphasizes the universal, timeless nature of romantic passion.

Kiss Quatrain with Soraismus

When soft lips meet mine with tender grace,
And corazóns beat wild beneath the stars,
Our souls unite in love's sweet embrace,
While moonlight heals all our hidden scars.

Note: This quatrain employs soraismus in "corazóns" which blends the Spanish "corazón" (heart) with an English plural ending "-s," creating a hybrid linguistic form. This mixing of Spanish and English morphology demonstrates soraismus by deliberately confusing linguistic boundaries, suggesting that passionate love creates its own language that transcends conventional grammatical rules.

Kiss Quatrain with Heterogenium

When soft lips meet mine with tender grace,
And hearts beat wild sotto the starlit sky,
Our souls unite in love's sweet embrace,
While moonbeams dance and gentle zéphyrs sigh.

Note: This quatrain employs heterogenium by mixing incongruous linguistic elements: "sotto" (Italian "under") inappropriately placed in English syntax, and "zéphyrs" with French accent marking on an English word. These mismatched linguistic borrowings create deliberate incongruity that suggests passionate love disrupts normal language patterns, mixing vocabularies in emotionally charged but grammatically inappropriate ways.

Kiss Quatrain with Amphibology

When soft lips meet mine with tender grace,
I kissed her standing by the starlit tree,
Our souls unite in love's sweet embrace,
While moonbeams dance in perfect harmony.

Note: This quatrain employs amphibology in "I kissed her standing by the starlit tree" where the syntax creates ambiguity about who is standing by the tree. The phrase could mean either "I kissed her while I was standing by the tree" or "I kissed her while she was standing by the tree." This deliberate grammatical ambiguity adds layers of interpretation to the romantic scene.

Kiss Quatrain with Tapinosis

When soft lips meet mine with tender grace,
A mere peck beneath the starlit sky,
Our souls unite in love's sweet embrace,
While moonbeams dance and gentle breezes sigh.

Note: This quatrain employs tapinosis in describing the kiss as "A mere peck" which deliberately diminishes what is clearly presented as a profound, soul-uniting romantic moment. This understated description creates ironic contrast with the elevated language surrounding it ("souls unite," "love's sweet embrace"), emphasizing through diminishment the actually transcendent nature of the kiss.

Kiss Quatrain with Cacemphaton

When soft lips meet mine with tender grace,
And hearts beat wild beneath the starlit sky,
Our souls unite in love's sweet embrace,
While moon beams dance and gentle breezes sigh.

Note: This quatrain employs cacemphaton in "While moon beams dance" where the separation of "moonbeams" into "moon beams" creates an awkward sound combination when read aloud, as "moon beams" can sound like "moon beans" or create an unnatural pause. This deliberate arrangement demonstrates how word positioning can create unintended sonic effects that might distract from or complicate the intended meaning.

Kiss Quatrain with Tautology

When soft lips meet mine with tender grace,
And hearts beat wild beneath the starlit sky,
Our souls unite together in one place,
While moonbeams dance and gentle breezes sigh.

Note: This quatrain employs tautology in "Our souls unite together in one place" where "unite" and "together in one place" express essentially the same concept of joining or coming together. This repetition of the same idea in different words creates emphasis on the profound connection between the lovers, reinforcing through redundancy the completeness of their romantic union during the kiss.

Kiss Quatrain with Perissologia

When soft and tender lips meet mine with grace,
And hearts beat wild and fast beneath the sky,
Our souls unite together in embrace,
While bright moonbeams dance and breezes gently sigh.

Note: This quatrain employs perissologia through excessive modification and redundancy: "soft and tender lips" (both adjectives mean essentially the same), "wild and fast" (redundant intensifiers), "unite together" (unite already implies togetherness), and "bright moonbeams" with "gently sigh" (unnecessary descriptors). This verbose style creates emphasis through abundance of language, though it risks diluting the impact through over-elaboration.

Kiss Quatrain with Macrologia

When soft lips meet mine with tender grace,
And cardiovascular organs pulsate wild,
Our souls unite in love's sweet embrace,
While moonbeams dance and breezes blow so mild.

Note: This quatrain employs macrologia in "cardiovascular organs pulsate" where the unnecessarily elaborate and clinical "cardiovascular organs pulsate" is used instead of the simple "hearts beat." This verbose, pompous phrasing creates an absurd contrast with the intimate romantic moment, demonstrating how overly elaborate language can sometimes undermine rather than enhance poetic expression.

Kiss Quatrain with Parelcon

When soft lips meet mine with tender grace,
And hearts beat wild beneath the starlit sky,
Our souls unite in love's sweet embrace there,
While moonbeams dance and gentle breezes sigh.

Note: This quatrain employs parelcon in "Our souls unite in love's sweet embrace there" where the word "there" is superfluous and adds nothing to the meaning. The location is already established by the context, making "there" an unnecessary addition that demonstrates how extra words can creep into verse without adding value, sometimes weakening the impact through verbal excess.

Kiss Quatrain with Pleonasmus

When soft lips meet mine with tender grace,
And hearts beat wild beneath the starlit sky,
We both together join in love's embrace,
While moonbeams dance and gentle breezes sigh.

Note: This quatrain employs pleonasmus in "We both together join" where "both" and "together" are redundant - if "we" join, it's already implied that both parties are involved, and "together" repeats this concept unnecessarily. This redundancy demonstrates how pleonasmus can create emphasis through repetition, though it may also weaken prose through unnecessary verbosity.

Kiss Quatrain with Homologia

"This kiss," she said, "might break our hearts in two."
"You're right," I whispered, "yet I cannot stay
Away from lips so soft, so sweet, so true—
Let broken hearts be love's sweet price to pay."

Note: This quatrain employs homologia in "You're right," where the speaker concedes to the beloved's concern that their kiss might cause heartbreak. Rather than arguing against her point, he agrees with her assessment but then proceeds anyway, acknowledging the validity of her concern while choosing love despite the risk. This creates dramatic tension through agreement rather than conflict.

Kiss Quatrain with Periergia

When soft lips, delicate as morning dew
Upon the petals of a virgin rose
In gardens where the ancient cedars grew,
Meet mine beneath where starlight softly glows.

Note: This quatrain employs periergia through excessive ornamentation and unnecessary elaboration in describing the lips. The extended metaphor "delicate as morning dew / Upon the petals of a virgin rose / In gardens where the ancient cedars grew" creates overwrought, verbose expression that belabors a simple comparison. This demonstrates how excessive decoration can sometimes overwhelm the central image rather than enhance it.

Kiss Quatrain with Bomphiologia

When soft lips meet mine with tender grace,
The cosmic firmament itself rejoices,
Our souls unite in love's triumphant space,
While universal harmonies sing voices.

Note: This quatrain employs bomphiologia through inflated, grandiose language that seems excessive for describing a simple kiss: "cosmic firmament itself rejoices," "love's triumphant space," and "universal harmonies sing voices." This bombastic expression creates an inappropriately grand scale for an intimate moment, demonstrating how pompous language can sometimes make romantic poetry seem overwrought rather than heartfelt.

Kiss Quatrain with Cacozelia

When soft lips meet mine with tender grace,
And hearts palpitate 'neath stelliferous dome,
Our souls coalesce in love's sweet embrace,
While zephyrs perambulate 'round our home.

Note: This quatrain employs cacozelia through pretentious, affected language: "palpitate 'neath stelliferous dome" (instead of "beat beneath the starry sky"), "coalesce" (instead of "unite"), and "zephyrs perambulate" (instead of "breezes blow"). This artificially elaborate diction creates forced elegance that seems contrived rather than natural, demonstrating how overly sophisticated language can make poetry feel artificial.

Kiss Quatrain with Cacosyntheton

When soft lips meet mine with tender grace,
And hearts beat wild beneath the starlit sky,
Harsh-sweet sharp-soft touch ignites our face,
While moonbeams dance and gentle breezes sigh.

Note: This quatrain employs cacosyntheton in "Harsh-sweet sharp-soft touch" where contradictory adjectives are awkwardly combined, creating discordant, unpleasant word arrangements. The jarring combination of opposing qualities ("harsh-sweet," "sharp-soft") creates an uncomfortable, clashing effect that demonstrates how poor word choice and arrangement can disrupt the flow and harmony of romantic verse.

Kiss Quatrain with Malapropism

When soft lips meet mine with tender grace,
And hearts beat wild beneath the starlit sky,
Our souls ignite in love's sweet interface,
While moonbeams dance and gentle breezes cry.

Note: This quatrain employs malapropism in "love's sweet interface" where "interface" (a technical term for connection between systems) is incorrectly used instead of "embrace" or "interlace." This creates an unintentionally humorous effect by applying computer terminology to romance, demonstrating how the wrong word choice can disrupt poetic tone and create comic rather than romantic effect.

Kiss Quatrain with Alliteration

When soft lips sweetly seek mine with grace,
And hearts beat wild beneath the starlit sky,
Our souls unite in love's tender embrace,
While whispered words and wistful winds reply.

Note: This quatrain employs alliteration in multiple instances: "soft lips sweetly seek" (repeated 's' sounds) and "whispered words and wistful winds" (repeated 'w' sounds). This repetition of initial consonant sounds creates musical, rhythmic effects that enhance the lyrical quality of the verse and make it more memorable and pleasing to recite aloud.

Kiss Quatrain with Paroemion

When soft lips meet mine with tender grace,
Sweet sighs suspend the silence of the night,
Silent stars seem to shine with silver light,
While moonbeams dance around our warm embrace.

Note: This quatrain employs paroemion in the successive phrases "Sweet sighs suspend," "Silent stars seem," and "shine with silver" where the repeated 's' sound begins multiple clauses and phrases in parallel structure. This creates a cascading alliterative effect that enhances the musical quality while maintaining syntactic parallelism, demonstrating how paroemion combines sound repetition with structural organization.

Kiss Quatrain with Anaphora

When soft lips meet mine with tender grace,
When hearts beat wild beneath the starlit sky,
When souls unite in love's sweet embrace,
When moonbeams dance, our spirits learn to fly.

Note: This quatrain employs anaphora through the repetition of "When" at the beginning of each line. This parallel structure creates rhythmic emphasis and builds momentum through repetition, suggesting the accumulation of romantic elements that all converge in the moment of the kiss. The repeated "When" clauses create a sense of simultaneity and inevitability in the romantic encounter.

Kiss Quatrain with Epistrophe

When soft lips meet mine, my heart finds peace,
And wild hearts beat beneath stars, finding peace,
Our souls unite in love's embrace of peace,
While moonbeams dance in gentle, lasting peace.

Note: This quatrain employs epistrophe through the repetition of "peace" at the end of each line. This parallel structure creates rhythmic emphasis and reinforces the central theme that the kiss brings tranquility and harmony to all aspects of the romantic moment. The repeated ending creates a sense of resolution and completeness, emphasizing how love brings ultimate serenity.

Kiss Quatrain with Symploce

When soft lips meet mine with tender grace,
When hearts beat wild in this sacred space,
When souls unite, time cannot erase
When love blooms bright in our embrace.

Note: This quatrain employs symploce by combining anaphora (repetition of "When" at the beginning of each line) with epistrophe-like repetition in the similar endings ("grace," "space," "erase," "embrace"). This dual repetition creates both rhythmic momentum through the repeated openings and harmonic closure through the rhyming endings, demonstrating how symploce can create complex musical effects in verse.

Kiss Quatrain with Epanalepsis

Love brings soft lips to mine with tender love,
And hearts beat wild beneath the starlit sky,
Our souls unite in passion's sacred grove,
While moonbeams dance where gentle breezes sigh.

Note: This quatrain employs epanalepsis in the first line "Love brings soft lips to mine with tender love" where the word "love" appears at both the beginning and end of the line. This circular repetition creates emphasis and completeness, suggesting that love is both the source and the destination of the romantic encounter, encircling the entire experience of the kiss.

Kiss Quatrain with Antimetabole

When soft lips meet mine with tender grace,
My heart finds you, and you find my heart,
Our souls unite in love's sweet embrace,
While moonbeams dance as day and night depart.

Note: This quatrain employs antimetabole in "My heart finds you, and you find my heart" where the key words "heart," "finds," and "you" are repeated in reverse order. This chiasmic structure creates balanced, symmetrical meaning that emphasizes the mutual nature of love and discovery in the kiss, showing how both lovers simultaneously seek and are found by each other.

Kiss Quatrain with Anadiplosis

When soft lips meet mine with tender grace,
Grace flows through hearts beneath the starlit sky,
Sky watches as our souls unite on high,
High above earthly cares, love finds its place.

Note: This quatrain employs anadiplosis by repeating the last word of each line as the first word of the following line: "grace/Grace," "sky/Sky," and "high/High." This creates a chain-like connection between lines that builds momentum and emphasizes the progression from physical touch to spiritual transcendence, with each element flowing naturally into the next in the romantic sequence.

Kiss Quatrain with Polyptoton

When soft lips meet mine with tender grace,
And hearts beat wild beneath the starlit sky,
Our souls unite in love's sweet, loving space,
While moonbeams dance as lovers gently sigh.

Note: This quatrain employs polyptoton through the repetition of words from the same root in different forms: "love's" (possessive noun), "loving" (adjective), and "lovers" (plural noun). This repetition of the root "love" in various grammatical forms creates emphasis and reinforces the central theme while demonstrating the versatility and pervasiveness of love throughout the romantic moment.

Kiss Quatrain with Diaphora

When soft lips meet mine with tender grace,
And hearts beat wild beneath the starlit sky,
Our souls unite in love's sweet, sacred space—
This space between us now begins to die.

Note: This quatrain employs diaphora through the repetition of "space" with different meanings: first as "love's sweet, sacred space" (referring to the emotional/spiritual realm of their connection), and then as "This space between us" (referring to the physical distance). This creates a contrast between the expanding spiritual connection and the diminishing physical separation during the kiss.

Kiss Quatrain with Ploce

When soft lips meet mine with tender grace,
And hearts beat wild beneath the starlit sky,
Our souls unite in love's sweet embrace—
Sweet beyond words, sweet beyond reason's cry.

Note: This quatrain employs ploce through the repetition of "sweet" with intervening words: first in "love's sweet embrace" and then repeated twice in "Sweet beyond words, sweet beyond reason's cry." The word returns with intensified meaning, emphasizing how the sweetness of the kiss exceeds both linguistic expression and rational understanding, building emotional intensity through the delayed repetition.

Kiss Quatrain with Diacope

When soft, so soft, lips meet mine with grace,
And hearts beat wild beneath the starlit sky,
Our souls unite in love's sweet embrace,
While moonbeams dance and gentle breezes sigh.

Note: This quatrain employs diacope in "soft, so soft" where the word "soft" is repeated with only the word "so" intervening. This immediate repetition creates emphasis and intensifies the sensual quality of the lips, demonstrating how diacope can heighten emotional impact through the close proximity of repeated words that build upon each other for greater effect.

Kiss Quatrain with Epizeuxis

When soft lips meet, meet mine with tender grace,
And hearts beat wild beneath the starlit sky,
Our souls unite in love's sweet embrace,
While moonbeams dance and gentle breezes sigh.

Note: This quatrain employs epizeuxis in "meet, meet mine" where the word "meet" is repeated immediately without intervening words. This creates intense emphasis and suggests the urgency or overwhelming nature of the moment when lips come together, demonstrating how epizeuxis can convey emotional intensity through the immediate, breathless repetition of key words.

Kiss Sonnet with Testimony

When soft lips met mine with tender grace,

The stars themselves bore witness to our love,

And moonbeams testified with silver face

That heaven smiled upon us from above.

The ancient poets spoke of such delight—

Their verses stand as proof of passion's power,

How Sappho sang of love's enchanting might,

And Ovid chronicled each stolen hour.

The very wind confessed what hearts had known,

The roses nodded, swearing to the truth

That in this kiss, two souls had fully grown

Beyond the simple boundaries of youth.

Let nature's testimony make it clear:

This kiss was blessed by all that we hold dear.

Note: This sonnet employs testimony through multiple forms of evidence: the stars "bore witness," moonbeams "testified," ancient poets like Sappho and Ovid provide literary precedent, the wind "confessed," and roses "swore to the truth." These various testimonies from nature, literature, and the elements all support the claim that this kiss was extraordinary and blessed, using the rhetorical device to build a case for love's transcendent power.

The sonnet uses testimony by presenting multiple forms of evidence to support the claim that the kiss was extraordinary: celestial witnesses (stars, moonbeams), literary authorities (Sappho, Ovid), and natural elements (wind, roses) all testify to the kiss's significance. This accumulation of testimonial evidence creates a compelling argument for love's transcendent power, demonstrating how the rhetorical device can elevate a personal moment to universal importance through authoritative validation.

Retry
N

add to the wisdom of literary expression by writing a poetic sonnet about a kiss. please be sure to use an example of the word schema known as martyria in the poem

Edit

I'll create a Shakespearean sonnet about a kiss that incorporates martyria, the rhetorical device where personal testimony or first-hand witness is offered as evidence, often emphasizing the speaker's direct experience.

Kiss Sonnet with Martyria
Document
Kiss Sonnet with Martyria
When soft lips met mine with tender grace,
I witnessed love transform the very air—
My own eyes saw how starlight touched your face,
My heart can testify beyond compare.
I felt the world grow still beneath our touch,
I h

The sonnet uses martyria by emphasizing the speaker's personal testimony through phrases like "I witnessed," "My own eyes saw," "My heart can testify," and "I stand as witness." This first-hand evidence creates powerful authenticity, as the speaker offers their direct experience as proof of the kiss's transformative power, demonstrating how personal testimony can serve as the most compelling evidence for love's profound effects on the human soul.

Retry
N

add to the wisdom of literary expression by writing a poetic sonnet about a kiss. please be sure to use an example of the word schema known as proverb in the poem

Edit

I'll create a Shakespearean sonnet about a kiss that incorporates a proverb, the rhetorical device where traditional wisdom or folk sayings are used to support or illustrate a point.

Kiss Sonnet with Proverb
Document

Kiss Sonnet with Proverb

When soft lips met mine with tender grace,
The world grew still beneath the starlit sky,
As actions spoke much louder than words' place—
Your kiss revealed what speech could not imply.
No need for vows or promises so swee

The sonnet uses the proverb "actions speak louder than words" (expressed as "actions spoke much louder than words' place") to support the argument that the kiss communicates more powerfully than any verbal declaration of love. This traditional wisdom reinforces the central theme that physical intimacy can convey emotions more effectively than language, demonstrating how proverbs can lend authoritative weight to romantic poetry through time-tested human understanding.

Retry
N

add to the wisdom of literary expression by writing a poetic sonnet about a kiss. please be sure to use an example of the word schema known as adage in the poem

Edit

I'll create a Shakespearean sonnet about a kiss that incorporates an adage, the rhetorical device where traditional sayings or time-honored expressions of wisdom are used to support or illuminate a point.

Kiss Sonnet with Adage
Document
Kiss Sonnet with Adage When soft lips met mine with tender grace, The world grew still beneath the starlit sky, And in that moment's sweet, eternal space, I learned what ages past could not deny. The wise have said that love makes fools of all, Y

The sonnet uses adages through the traditional sayings "love makes fools of all" and "all good things must end," employing these time-honored expressions as conventional wisdom that the extraordinary kiss challenges and transcends. This demonstrates how adages can serve as rhetorical touchstones against which exceptional experiences are measured, showing how true love can defy even the most accepted truths about human nature and mortality.

Retry
N

add to the wisdom of literary expression by writing a poetic sonnet about a kiss. please be sure to use an example of the word schema known as apothegm in the poem

Edit

I'll create a Shakespearean sonnet about a kiss that incorporates an apothegm, the rhetorical device where a terse, witty, or memorable saying is used to convey wisdom or insight.

Kiss Sonnet with Apothegm
Document
Kiss Sonnet with Apothegm When soft lips met mine with tender grace, The world grew still beneath the starlit sky, And in that sweet, transcendent embrace, I learned a truth that words cannot deny. The poets speak of love in lengthy verse, Philos

The sonnet uses the apothegm "The shortest distance runs from lip to heart," a terse, witty adaptation of geometric wisdom applied to romance. This memorable saying captures the direct connection between physical intimacy and emotional truth, demonstrating how apothegms can distill complex insights into pointed, memorable expressions that convey profound wisdom about love's immediate and transformative power through concise, clever phrasing.

Retry
N

add to the wisdom of literary expression by writing a poetic sonnet about a kiss. please be sure to use an example of the word schema known as maxim in the poem

Edit

I'll create a Shakespearean sonnet about a kiss that incorporates a maxim, the rhetorical device where a general principle or rule of conduct is expressed as a concise statement of fundamental truth.

Kiss Sonnet with Maxim
Document
Kiss Sonnet with Maxim When soft lips met mine with tender grace, The world grew still beneath the starlit sky, And in that sweet, eternal embrace, I learned what truth could never tell a lie. The sages teach that silence speaks the most, And in

The sonnet uses maxims through fundamental principles like "silence speaks the most," "actions are the truest form of prayer," and "True hearts speak clearest when they touch, not planned." These concise statements of general truth about communication, spirituality, and love support the poem's central argument that the kiss transcends verbal expression, demonstrating how maxims can provide authoritative foundation for romantic philosophy through universal principles.

Retry
N

add to the wisdom of literary expression by writing a poetic sonnet about a kiss. please be sure to use an example of the word schema known as sententious in the poem

Edit

I'll create a Shakespearean sonnet about a kiss that incorporates sententious language, the rhetorical device where morally instructive, wise, or aphoristic statements are used, often with a didactic or moralizing tone.

Kiss Sonnet with Sententious Language
Document
Kiss Sonnet with Sententious Language
Document · Version 2
Kiss Sonnet with Sententious Language When soft lips met mine with tender grace, The world grew still beneath the starlit sky, And wisdom taught me in that sweet embrace: True virtue lies not in what we deny. He who would know love's purest, holi

The sonnet uses sententious language through morally instructive statements like "True virtue lies not in what we deny" and "He who would know love's purest, holiest art / Must first surrender pride's defensive wall." These didactic, aphoristic expressions present the kiss as a moral lesson about courage, humility, and virtue, demonstrating how sententious rhetoric can transform a romantic moment into wise instruction about love's deeper spiritual and ethical significance.

Retry
N

add to the wisdom of literary expression by writing a poetic sonnet about a kiss. please be sure to use an example of the word schema or figure known as diatyposis in the poem

Edit

I'll create a Shakespearean sonnet about a kiss that incorporates diatyposis, the rhetorical figure where vivid, detailed description creates a clear mental picture or impression through graphic portrayal.

Kiss Sonnet with Diatyposis
Document
Kiss Sonnet with Diatyposis When soft lips met mine with tender grace, I felt the velvet warmth of rose petals Press gently to my mouth, while your sweet face Glowed amber in the moonlight's silver metals. Your breath, like honeyed wine, caressed

The sonnet uses diatyposis through vivid, detailed imagery like "velvet warmth of rose petals," "amber in moonlight's silver metals," "honeyed wine," and "twilight's purple gloom." These graphic, sensory descriptions create a complete mental picture that allows readers to visualize and almost experience the kiss themselves, demonstrating how diatyposis can transform abstract emotion into tangible, memorable imagery through detailed portrayal of physical sensations and atmospheric details.

Retry
N

add to the wisdom of literary expression by writing a poetic sonnet about a kiss. please be sure to use an example of the word schema or figure known as apomnemonysis in the poem

Edit

I'll create a Shakespearean sonnet about a kiss that incorporates apomnemonysis, the rhetorical figure where past events are recalled or cited as examples to support a present argument or situation.

Kiss Sonnet with Apomnemonysis
Document
Kiss Sonnet with Apomnemonysis When soft lips met mine with tender grace, I recalled how Paris stole fair Helen's heart, How Cleopatra's kiss sealed Caesar's fate, And Romeo found his soul's immortal part. Remember how the poets sang of old— When

The sonnet uses apomnemonysis by recalling famous historical and literary kisses—Paris and Helen, Cleopatra and Caesar, Romeo and Juliet, Sappho's love, Dido's passion, and Hero and Leander. These past examples are cited to provide context and comparison, ultimately arguing that the present kiss surpasses all these celebrated loves from history and legend, demonstrating how citing past events can enhance the significance of a current experience through comparative testimony.

Retry
N

add to the wisdom of literary expression by writing a poetic sonnet about a kiss. please be sure to use an example of the word schema or figure known as epicrisis in the poem

Edit

I'll create a Shakespearean sonnet about a kiss that incorporates epicrisis, the rhetorical figure where a judgment or critical evaluation is made, often weighing different aspects or making a decisive assessment.

Kiss Sonnet with Epicrisis
Document
Kiss Sonnet with Epicrisis When soft lips met mine with tender grace, The world grew still beneath the starlit sky, And in that sweet, transcendent embrace, I weighed what poets claim and what souls cry. Some say that kisses are but fleeting thin

The sonnet uses epicrisis through the speaker's deliberate judgment and critical evaluation: "I weighed what poets claim and what souls cry," "I judge between these two opposing wings," and "My final verdict on love's sweetest art." This demonstrates how the rhetorical figure involves careful assessment of different perspectives before rendering a decisive conclusion, showing the kiss as both temporal and eternal, balancing opposing viewpoints through reasoned evaluation.

Retry
N

add to the wisdom of literary expression by writing a poetic sonnet about a kiss. please be sure to use an example of the word schema or figure known as chria in the poem

Edit

I'll create a Shakespearean sonnet about a kiss that incorporates chria, the rhetorical figure where a brief anecdote or saying is attributed to a specific person and then elaborated upon or applied to the current situation.

Kiss Sonnet with Chria
Document
Kiss Sonnet with Chria When soft lips met mine with tender grace, I thought of what old Socrates once said: "The only thing I know is that I know Nothing at all"—yet here, beside your face, His wisdom crumbles like the morning's snow. For in this

The sonnet uses chria by citing Socrates' famous saying "The only thing I know is that I know nothing at all" and then elaborating on how this philosophical principle applies to the kiss experience. The speaker contrasts

Socrates' admission of ignorance with the certain knowledge gained through love's touch, demonstrating how chria allows poets to use attributed wisdom as a springboard for exploring how ancient insights relate to present romantic revelation.

Retry
N

add to the wisdom of literary expression by writing a poetic sonnet about a kiss. please be sure to use an example of the word schema or figure known as an oath in the poem

Edit

I'll create a Shakespearean sonnet about a kiss that incorporates an oath, the rhetorical figure where a solemn promise or vow is made, often invoking divine or sacred authority as witness.

Kiss Sonnet with Oath
Document
Kiss Sonnet with Oath When soft lips met mine with tender grace, The world grew still beneath the starlit sky, And in that sweet, transcendent embrace, I swore by all the stars that will not die: By heaven's vault and earth's eternal ground, By s

The sonnet uses an oath by invoking sacred witnesses and making solemn vows: "I swore by all the stars that will not die," "By heaven's vault and earth's eternal ground," "I vow by love itself," and "So help me God." These sacred invocations demonstrate how oaths can elevate a romantic moment to the level of divine commitment, showing the transformative power of the kiss through the speaker's willingness to make binding promises witnessed by the highest authorities.

Retry
N

add to the wisdom of literary expression by writing a poetic sonnet about a kiss. please be sure to use an example of the word schema or figure known as a vow in the poem

Edit

I'll create a Shakespearean sonnet about a kiss that incorporates a vow, the rhetorical figure where a solemn promise or pledge is made, often expressing commitment or dedication to a course of action or belief.

Kiss Sonnet with Vow
Document
Kiss Sonnet with Vow When soft lips met mine with tender grace, The world grew still beneath the starlit sky, And in that sweet, transcendent embrace, I made a promise that shall never die: I vow to hold this moment ever near, To treasure every b

The sonnet uses a vow through solemn promises like "I vow to hold this moment ever near," "I pledge my heart to memory's sacred trust," and "I promise you, my soul, my very life." These commitments transform the romantic moment into binding dedication, demonstrating how vows can elevate a kiss from a fleeting physical act to a sacred covenant, showing the speaker's determination to honor and preserve the kiss's significance throughout their entire lifetime.

Retry
N

add to the wisdom of literary expression by writing a poetic sonnet about a kiss. please be sure to use an example of the word schema or figure known as a pledge in the poem

Edit

I'll create a Shakespearean sonnet about a kiss that incorporates a pledge, the rhetorical figure where a formal promise or commitment is made, often as security or guarantee of one's intentions.

Kiss Sonnet with Pledge
Document
Kiss Sonnet with Pledge When soft lips met mine with tender grace, The world grew still beneath the starlit sky, And in that sweet, transcendent embrace, I offered up my heart's most sacred tie: I pledge this kiss as token of my truth, A seal upo

The sonnet uses a pledge through formal commitments like "I pledge this kiss as token of my truth," "Let this soft touch stand surety for my word," and "My heart, my soul, my life—I pledge to you." These declarations transform the kiss into a binding guarantee, demonstrating how pledges can elevate a romantic moment into a formal contract of love, where the physical act serves as security for future devotion and unwavering commitment.

Retry
N

add to the wisdom of literary expression by writing a poetic sonnet about a kiss. please be sure to use an example of the word schema or figure known as euche in the poem

Edit

I'll create a Shakespearean sonnet about a kiss that incorporates euche, the rhetorical figure where a prayer, wish, or earnest petition is made, often expressing hope or desire for divine favor or blessing.

Kiss Sonnet with Euche
Document
Kiss Sonnet with Euche When soft lips met mine with tender grace, The world grew still beneath the starlit sky, And in that sweet, transcendent embrace, I lifted up my heart to heaven high: O grant, kind Fortune, that this kiss may last Beyond th

The sonnet uses euche through prayers and earnest petitions like "O grant, kind Fortune," "May angels guard," "I pray that time may never dim," and "May heaven smile upon." These prayerful expressions demonstrate how euche can transform romantic experience into spiritual petition, seeking divine blessing and protection for the love inspired by the kiss, showing how profound emotion naturally turns toward prayer and hope for heavenly favor.

Retry
N

add to the wisdom of literary expression by writing a poetic sonnet about a kiss. please be sure to use an example of the word schema or figure known as eustathia in the poem

Edit

I'll create a Shakespearean sonnet about a kiss that incorporates eustathia, the rhetorical figure where stability, steadfastness, or unwavering constancy is emphasized, often expressing firm resolve or unchanging commitment.

Kiss Sonnet with Eustathia
Document
Kiss Sonnet with Eustathia When soft lips met mine with tender grace, The world grew still beneath the starlit sky, And in that sweet, transcendent embrace, I found

what time can never modify: Though seasons change and years may pass away, Though

The sonnet uses eustathia through expressions of unwavering constancy: "I found what time can never modify," "This kiss shall stand unmoved," "A constant star," "No earthly power can shake," and "Steadfast and sure." These declarations emphasize the unchanging, stable nature of the love inspired by the kiss, demonstrating how eustathia can express firm resolve and unchanging commitment that remains constant despite all temporal changes and external challenges.

Retry
N

add to the wisdom of literary expression by writing a poetic sonnet about a kiss. please be sure to use an example of the word schema or figure known as asphalia in the poem

Edit

I'll create a Shakespearean sonnet about a kiss that incorporates asphalia, the rhetorical figure where security, safety, or assurance is emphasized, often expressing confidence in protection or certainty about outcomes.

Kiss Sonnet with Asphalia
Document

Kiss Sonnet with Asphalia When soft lips met mine with tender grace, The world grew still beneath the starlit sky, And in that sweet, transcendent embrace, I found the safety that will never die: This kiss has made me certain of my ground, Secure

The sonnet uses asphalia through expressions of security and assurance: "I found the safety that will never die," "certain of my ground," "Secure within love's fortress," "sheltered from all harm," "Protected by the vow," and "A sanctuary where my soul can live." These declarations emphasize the protective, secure nature of the love established by the kiss, demonstrating how asphalia can express confidence in love's safeguarding power and the certainty of emotional protection.

Retry
N

add to the wisdom of literary expression by writing a poetic sonnet about a kiss. please be sure to use an example of the word schema or figure known as euphemismus in the poem

Edit

I'll create a Shakespearean sonnet about a kiss that incorporates euphemismus, the rhetorical figure where mild, indirect, or pleasant terms are used to replace harsh, direct, or unpleasant expressions.

Kiss Sonnet with Euphemismus

Document
Kiss Sonnet with Euphemismus When soft lips met mine with tender grace, The world grew still beneath the starlit sky, And in that sweet, transcendent embrace, We shared love's gentle dance with passion's sigh. Though others might speak crudely of

The sonnet uses euphemismus by employing gentle, elevated language to describe physical passion: "love's gentle dance with passion's sigh," "sweet union of the heart," "souls communed," and "tasted heaven's nectar" instead of more direct or potentially crude descriptions of romantic desire. These mild, pleasant terms demonstrate how euphemismus can elevate and refine discussion of intimate moments, transforming potentially harsh realities into beautiful, spiritual expressions.

Retry
Claude can make mistakes.
Please double-check responses.
This conversation has reached its maximum length.
Start a new conversation

Sonnet 4

Kiss Sonnet with Martyria

When soft lips met mine with tender grace,
I witnessed love transform the very air—
My own eyes saw how starlight touched your face,
My heart can testify beyond compare.
I felt the world grow still beneath our touch,
I heard the silence sing a sacred song,
I bore witness to a moment meant as such
That time itself confessed it lingered long.
These lips have known no sweeter, softer kiss,
These hands have never trembled quite this way,
I stand as witness to this perfect bliss—
My soul attests what happened on this day.
Let others doubt what passion's power can do:
I swear by all I've felt—this love is true.

Note: This sonnet employs martyria through the speaker's repeated personal testimony using phrases like "I witnessed," "My own eyes saw," "My heart can testify," "I felt," "I heard," "I bore witness," and "I stand as witness." The speaker offers their direct, first-hand experience as evidence of the kiss's transformative power, emphasizing personal observation and feeling as authoritative proof of love's reality.

Kiss Sonnet with Proverb

When soft lips met mine with tender grace,

The world grew still beneath the starlit sky,

As actions spoke much louder than words' place—

Your kiss revealed what speech could not imply.

No need for vows or promises so sweet,

No lengthy verse to capture love's true art,

For deeds, not words, make passion's truth complete,

And one soft touch can speak to any heart.

The ancient wisdom rings forever true:

What lips can do, no language can convey,

A single kiss says more than "I love you"

Could ever hope to capture or portray.

Let poets pen their verses, line by line—

Your kiss spoke volumes, silent and divine.

Note: This sonnet employs a proverb through the variation of "actions speak louder than words," presenting it as "actions spoke much louder than words' place" and reinforcing it with "deeds, not words, make passion's truth complete." The sonnet uses this traditional wisdom to argue that the kiss communicates more effectively than any verbal declaration, demonstrating how proverbs can provide authoritative support for romantic arguments.

Kiss Sonnet with Adage

When soft lips met mine with tender grace,

The world grew still beneath the starlit sky,

And in that moment's sweet, eternal space,

I learned what ages past could not deny.

The wise have said that love makes fools of all,

Yet in your kiss I found a different truth—

That folly's wisdom answers passion's call,

And age finds meaning in the heart of youth.

For though they say that all good things must end,

This kiss defied time's ancient, cruel decree,

And proved that hearts in love can transcend

The boundaries of mortality.

Let sages quote their truths both old and new—

This kiss taught wisdom only love can brew.

Note: This sonnet employs adages through traditional sayings: "love makes fools of all" and "all good things must end." These time-honored expressions provide conventional wisdom that the speaker both acknowledges and challenges, using the kiss as evidence that transcends ordinary understanding. The adages serve as foils against which the extraordinary nature of this particular romantic moment can be measured and found exceptional.

Kiss Sonnet with Apothegm

When soft lips met mine with tender grace,
The world grew still beneath the starlit sky,
And in that sweet, transcendent embrace,
I learned a truth that words cannot deny.
The poets speak of love in lengthy verse,
Philosophy debates the heart's true art,
But wisdom came to me both swift and terse:
The shortest distance runs from lip to heart.
No lengthy discourse, no elaborate song
Could capture what that single moment bore—
Sometimes the briefest touch says more than long
Soliloquies of passion's ancient lore.
Let scholars pen their treatises above—
One kiss contains the very soul of love.

Note: This sonnet employs an apothegm in "The shortest distance runs from lip to heart," which is a terse, memorable adaptation of the geometric principle "the shortest distance between two points is a straight line." This witty, condensed wisdom captures the direct, immediate connection between physical intimacy and emotional truth, demonstrating how apothegms can distill complex romantic insights into memorable, pointed expressions.

Kiss Sonnet with Maxim

When soft lips met mine with tender grace,

The world grew still beneath the starlit sky,

And in that sweet, eternal embrace,

I learned what truth could never tell a lie.

The sages teach that silence speaks the most,

And in your kiss this wisdom came to light—

No words could match what tender lips can boast,

No speech surpass love's wordless, pure delight.

For actions are the truest form of prayer,

And every kiss becomes a sacred vow,

A testament beyond what tongues declare,

A promise sealed in passion's holy now.

Let those who doubt love's power understand:

True hearts speak clearest when they touch, not planned.

Note: This sonnet employs maxims through several concise statements of fundamental truth: "silence speaks the most," "actions are the truest form of prayer," and "True hearts speak clearest when they touch, not planned." These maxims serve as general principles about communication, spirituality, and love that support the poem's argument about the kiss's profound significance beyond mere words.

Kiss Sonnet with Sentimentious Language

When soft lips met mine with tender grace,
The world grew still beneath the starlit sky,
And wisdom taught me in that sweet embrace:
True virtue lies not in what we deny.
He who would know love's purest, holiest art
Must first surrender pride's defensive wall,
For only humble souls can truly start
To learn what passion's greatest teachers call.
The prudent heart that guards itself too well
Shall never taste the nectar of the soul,
But those who dare love's sacred truth to tell
Find in one kiss what makes the broken whole.
Let this be known to all who seek love's way:
The brave who kiss shall conquer, not betray.

Note: This sonnet employs sententious language through morally instructive statements like "True virtue lies not in what we deny," "He who would know love's purest, holiest art / Must first surrender pride's defensive wall," and "Let this be known to all who seek love's way." These didactic, aphoristic expressions present the kiss as a moral lesson about courage, humility, and the virtue found in authentic emotional expression.

Kiss Sonnet with Diatyposis

When soft lips met mine with tender grace,

I felt the velvet warmth of rose petals

Press gently to my mouth, while your sweet face

Glowed amber in the moonlight's silver metals.

Your breath, like honeyed wine, caressed my skin,

Your eyes half-closed in passion's drowsy dream,

While fingertips traced soft my trembling chin

And starlight caught love's crystalline gleam.

The air grew thick with jasmine's perfume

As pulse met pulse in rhythm's sacred dance,

Our shadows merged in twilight's purple gloom

While time stood still in love's entranced trance.

Such vivid beauty burned into my soul—

That single kiss made broken spirits whole.

Note: This sonnet employs diatyposis through detailed, vivid imagery that creates a clear mental picture: "velvet warmth of rose petals," "amber in moonlight's silver metals," "honeyed wine," "crystalline gleam," "jasmine's perfume," and "twilight's purple gloom." These graphic, sensory descriptions paint a complete, vivid scene that allows readers to visualize and almost experience the kiss through detailed portrayal.

Kiss Sonnet with Apomnemonysis

When soft lips met mine with tender grace,
I recalled how Paris stole fair Helen's heart,
How Cleopatra's kiss sealed Caesar's fate,
And Romeo found his soul's immortal part.
Remember how the poets sang of old—
When Sappho felt her lover's burning touch,
When Dido's passion made her spirit bold,
When Hero waited, loving Leander much.
These ancient tales of kisses lost in time
Pale next to what this moment brought to me,
For history's greatest loves, though deemed sublime,
Were shadows of this sweet reality.
Let scholars cite what legends may have done—
This kiss surpassed them all, and I'm the one.

Note: This sonnet employs apomnemonysis by recalling multiple historical and legendary examples of famous kisses and lovers: Paris and Helen, Cleopatra and Caesar, Romeo and Juliet, Sappho, Dido, and Hero and Leander. These past events are cited to provide context and comparison, ultimately arguing that the present kiss surpasses all these celebrated examples from history and literature.

Kiss Sonnet with Epicrisis

When soft lips met mine with tender grace,
The world grew still beneath the starlit sky,
And in that sweet, transcendent embrace,
I weighed what poets claim and what souls cry.
Some say that kisses are but fleeting things,
While others claim they hold eternal power—
I judge between these two opposing wings
And find the truth lies in this very hour.
For though this kiss must end as all things do,
Its memory shall burn beyond all time,
Neither purely mortal nor divine true,
But balanced perfectly 'tween earth and rhyme.
My final verdict on love's sweetest art:
This kiss was both—a moment and a heart.

Note: This sonnet employs epicrisis through the speaker's deliberate judgment and evaluation of opposing views about kisses. The lines "I weighed what poets claim and what souls cry," "I judge between these two opposing wings," and "My final verdict on love's sweetest art" demonstrate critical assessment, where the speaker considers different perspectives and renders a decisive evaluation about the nature and significance of the kiss.

Kiss Sonnet with Chria

When soft lips met mine with tender grace,

I thought of what old Socrates once said:

"The only thing I know is that I know

Nothing at all"—yet here, beside your face,

His wisdom crumbles like the morning's snow.

For in this kiss, all knowledge finds its thread,

All mysteries resolve in passion's glow,

All questions answered by love's gentle tread.

The philosopher who claimed he knew not truth

Would marvel at this moment's clarity—

How one sweet touch could teach what years of youth

Spent seeking wisdom could not guarantee.

If Socrates had tasted love like this,

He'd know that knowledge lives within a kiss.

Note: This sonnet employs chria by attributing the famous saying "The only thing I know is that I know nothing at all" to Socrates, then elaborating on how this philosophical principle applies to the present situation. The speaker uses Socrates' admission of ignorance to contrast with the certain knowledge gained through the kiss, demonstrating how the ancient wisdom relates to and is transformed by the current romantic experience.

Kiss Sonnet with Oath

When soft lips met mine with tender grace,
The world grew still beneath the starlit sky,
And in that sweet, transcendent embrace,
I swore by all the stars that will not die:
By heaven's vault and earth's eternal ground,
By sun and moon that mark our days and years,
By every sacred thing that can be found,
I pledge this kiss shall banish all my fears.
I vow by love itself, by passion's fire,
By all that's holy in this mortal realm,
That what this moment kindled of desire
Shall be the light by which my soul's at helm.
So help me God, and let the angels sing—
This kiss shall be my heart's eternal spring.

Note: This sonnet employs an oath through solemn vows invoking divine and natural authorities as witnesses: "I swore by all the stars," "By heaven's vault and earth's eternal ground," "By sun and moon," "I vow by love itself," and "So help me God." These sacred invocations demonstrate the rhetorical power of oaths to emphasize the profound significance and binding nature of the romantic commitment inspired by the kiss.

Kiss Sonnet with Vow

When soft lips met mine with tender grace,
The world grew still beneath the starlit sky,
And in that sweet, transcendent embrace,
I made a promise that shall never die:
I vow to hold this moment ever near,
To treasure every breath we shared as one,
To keep alive what started tender here,
Until my mortal days on earth are done.
I pledge my heart to memory's sacred trust,
That time shall never dim what love has wrought,
That though our bodies crumble into dust,
This kiss shall live in every loving thought.
I promise you, my soul, my very life—
This kiss shall cut through sorrow like a knife.

Note: This sonnet employs a vow through solemn promises and pledges: "I made a promise that shall never die," "I vow to hold this moment ever near," "I pledge my heart to memory's sacred trust," and "I promise you, my soul, my very life." These commitments demonstrate how vows can transform a romantic moment into a binding dedication, showing the speaker's determination to honor and preserve the significance of the kiss throughout their lifetime.

Kiss Sonnet with Pledge

When soft lips met mine with tender grace,

The world grew still beneath the starlit sky,

And in that sweet, transcendent embrace,

I offered up my heart's most sacred tie:

I pledge this kiss as token of my truth,

A seal upon the covenant we make,

My bond that through the trials beyond youth,

This love shall endure for both our sake.

Let this soft touch stand surety for my word,

A guarantee that what we've started here

Shall not be broken, lost, or left unheard—

This kiss, my pledge, shall cast away all fear.

Take this my promise, sealed with lips so true:

My heart, my soul, my life—I pledge to you.

Note: This sonnet employs a pledge through formal commitments and promises: "I pledge this kiss as token of my truth," "A seal upon the covenant we make," "My bond that through the trials," "Let this soft touch stand surety for my word," and "My heart, my soul, my life—I pledge to you." These formal declarations demonstrate how a pledge transforms the kiss into a binding guarantee, offering it as security for future devotion and commitment.

Kiss Sonnet with Euche

When soft lips met mine with tender grace,

The world grew still beneath the starlit sky,

And in that sweet, transcendent embrace,

I lifted up my heart to heaven high:

O grant, kind Fortune, that this kiss may last

Beyond the fleeting moment's swift decay,

May angels guard the love we've held so fast,

And bless this bond throughout each coming day.

I pray that time may never dim the fire

That sparked between us in this sacred hour,

O let the gods fulfill my heart's desire—

To keep forever love's transforming power.

May heaven smile upon what we have done,

And make these two hearts beat as if they're one.

Note: This sonnet employs euche through prayers and earnest petitions: "O grant, kind Fortune," "May angels guard," "I pray that time may never dim," "O let the gods fulfill my heart's desire," and "May heaven smile upon." These prayerful expressions demonstrate how euche can transform romantic experience into spiritual petition, seeking divine blessing and protection for the love kindled by the kiss.

Kiss Sonnet with Eustathia

When soft lips met mine with tender grace,

The world grew still beneath the starlit sky,

And in that sweet, transcendent embrace,

I found what time can never modify:

Though seasons change and years may pass away,

Though storms may rage and fortunes rise and fall,

This kiss shall stand unmoved from this day,

A constant star that guides me through it all.

No shifting wind can blow this love apart,

No earthly power can shake what we have built—

This moment lives forever in my heart,

Steadfast and sure, without regret or guilt.

Let all things change—I shall remain the same,

Forever marked by love's eternal flame.

Note: This sonnet employs eustathia through expressions of unwavering constancy and stability: "I found what time can never modify," "This kiss shall stand unmoved," "A constant star," "No shifting wind can blow this love apart," "No earthly power can shake," "Steadfast and sure," and "I shall remain the same." These declarations emphasize the unchanging, stable nature of the love inspired by the kiss, demonstrating firm resolve against all temporal changes.

Kiss Sonnet with Asphalia

When soft lips met mine with tender grace,
The world grew still beneath the starlit sky,
And in that sweet, transcendent embrace,
I found the safety that will never die:
This kiss has made me certain of my ground,
Secure within love's fortress, strong and true,
No fear can shake what here I have found—
A haven built for me alone by you.
In your arms I am sheltered from all harm,
Protected by the vow your lips have sealed,
Your love surrounds me like a gentle charm
That makes me safe from all the world's revealed.
This kiss ensures what nothing else can give:
A sanctuary where my soul can live.

Note: This sonnet employs asphalia through expressions of security and assurance: "I found the safety that will never die," "certain of my ground," "Secure within love's fortress," "No fear can shake," "sheltered from all harm," "Protected by the vow," "safe from all the world's," and "A sanctuary." These declarations emphasize the protective, secure nature of the love established by the kiss, demonstrating confidence in its safeguarding power.

Kiss Sonnet with Euphemismus

When soft lips met mine with tender grace,
The world grew still beneath the starlit sky,
And in that sweet, transcendent embrace,
We shared love's gentle dance with passion's sigh.
Though others might speak crudely of desire,
I'll say we found sweet union of the heart,
Our souls communed beside love's sacred fire,
As tender spirits chose to never part.
What vulgar tongues might call mere lust's demand,
I name the joining of two kindred souls,
A blessed meeting, pure and softly planned,
Where love's sweet mystery gently unfolds.
Let coarser minds reduce this to the flesh—
We tasted heaven's nectar, pure and fresh.

Note: This sonnet employs euphemismus by using gentle, elevated language to describe physical passion: "love's gentle dance with passion's sigh" instead of raw desire, "sweet union of the heart" and "souls communed" instead of physical attraction, "tender spirits" instead of bodily impulses, and "tasted heaven's nectar" instead of physical pleasure. These mild, pleasant terms replace potentially harsh or crude descriptions of romantic passion.

add to the wisdom of literary expression by writing a poetic quatrain about a kiss. please be sure to use an example of the word schema or figure known as euphemismus in the poem

Here's a quatrain about a kiss that employs euphemism:

A Quatrain on a Kiss

Document

The euphemism appears in "found sweet rest" and "tender congress" - gentler, more elevated expressions that replace direct descriptions of kissing and intimate connection. The phrase "sweet rest" transforms the physical act into something peaceful and restorative, while "tender congress" dignifies the meeting of lips as a formal assembly of souls rather than mere physical contact.

add to the wisdom of literary expression by writing a poetic sonnet about a kiss. please be sure to use an example of the word schema or figure known as euphemismus in the poem and add the name of the figure or schema into the title like, "kiss quatrain with euphemismus"

Kiss Sonnet with Euphemismus

Document · Version 2

This Shakespearean sonnet employs euphemismus throughout to elevate and refine the description of a kiss. The euphemistic expressions include "sweet communion" (instead of kissing), "tender parley" (gentle conversation between lips), "soft congress" (intimate meeting), "love's gentle warfare" (the passionate exchange), and "love's most sacred rite" (the kiss as ceremony). These refined substitutions transform the physical act into something spiritual and transcendent, demonstrating how euphemismus can add dignity and poetic beauty to even the most common human experiences.

add to the wisdom of literary expression by writing a poetic quatrain about a kiss. please be sure to use an example of the word schema or figure known as euphemismus in the poem and add the name of the figure or schema into the title like, "kiss quatrain with euphemismus"

Kiss Quatrain with Euphemismus

Document · Version 3

This quatrain employs euphemismus through "found sweet rest" and "tender congress" - refined expressions that dignify the kiss by replacing direct description with more elevated language. "Sweet rest" transforms the physical act into something peaceful and restorative, while "tender congress" presents the meeting of lips as a formal, almost parliamentary gathering of souls rather than mere physical contact.

add to the wisdom of literary expression by writing a poem with 6 to 8 lines about a kiss. please be sure to use an example of the word schema or figure known as euphemismus in the poem and add the name of the figure or schema (paranthetically) into the title like, "kiss bliss (with euphemismus)"

Kiss Bliss (with Euphemismus)

Document · Version 4

This six-line poem uses euphemismus in "gentle parley" (refined diplomatic conversation instead of kissing), "sweet retreat" (withdrawal into intimacy), and "tender congress" (soft assembly or meeting). These elevated expressions transform the physical act of kissing into something more ceremonial and spiritual, demonstrating how euphemismus can add grace and dignity to descriptions of intimate human moments.

Upon her lips I found sweet rest, Where honeyed whispers gently dwell, Two souls in tender congress blessed, Their secret language none can tell.

Upon her lips I found sweet rest,
Where honeyed whispers gently dwell,
Two souls in tender congress blessed,

Their secret language none can tell.
Kiss Sonnet with Euphemismus
Document · Version 2
When first our lips in sweet communion met, The world grew still as morning's gentle breath, Two hearts in tender parley, souls beget A language known to neither life nor death. Her mouth, a rose where honeyed secrets dwell, Received my prayer with

This Shakespearean sonnet employs euphemismus throughout to elevate and refine the description of a kiss. The euphemistic expressions include "sweet communion" (instead of kissing), "tender parley" (gentle conversation between lips), "soft congress" (intimate meeting), "love's gentle warfare" (the passionate exchange), and "love's most sacred rite" (the kiss as ceremony). These refined substitutions transform the physical act into something spiritual and transcendent, demonstrating how euphemismus can add dignity and poetic beauty to even the most common human experiences.

When first our lips in sweet communion met,
The world grew still as morning's gentle breath,
Two hearts in tender parley, souls beget A
language known to neither life nor death.
Her mouth, a rose where honeyed secrets dwell,
Received my prayer with grace beyond
compare, While whispered vows that only
lovers tell Were sealcd beneath the bower of her
hair.

In that soft congress of our yearning souls,
Time paused to witness love's most sacred rite,
As passion's fire through willing vessels rolls,
Consuming darkness, kindling heaven's light.

Thus in love's gentle warfare we found peace,
Where tender mercies never know surcease.

When first our lips in sweet communion met,
The world grew still as morning's gentle breath,
Two hearts in tender parley, souls beget
A language known to neither life nor death.

Her mouth, a rose where honeyed secrets dwell,
Received my prayer with grace beyond compare,
While whispered vows that only lovers tell
Were sealed beneath the bower of her hair.

In that soft congress of our yearning souls,
Time paused to witness love's most sacred rite,
As passion's fire through willing vessels rolls,
Consuming darkness, kindling heaven's light.

Thus in love's gentle warfare we found peace,
Where tender mercies never know surcease.

Kiss Quatrain with Euphemismus

Document · Version 3

Upon her lips I found sweet rest,
Where honeyed whispers gently dwell,
Two souls in tender congress blessed,
Their secret language none can tell.

This quatrain employs euphemismus through "found sweet rest" and "tender congress" - refined expressions that dignify the kiss by replacing direct description with more elevated language. "Sweet rest" transforms the physical act into something peaceful and restorative, while "tender congress" presents the meeting of lips as a formal, almost parliamentary gathering of souls rather than mere physical contact.

When first our lips in sweet communion met,
The world grew still as morning's gentle breath,
Two hearts in tender parley, souls beget
A language known to neither life nor death.
Her mouth, a rose where honeyed secrets dwell,
Received my prayer with grace beyond compare,
While whispered vows that only lovers tell
Were sealed beneath the bower of her hair.
In that soft congress of our yearning souls,
Time paused to witness love's most sacred rite,
As passion's fire through willing vessels rolls,
Consuming darkness, kindling heaven's light.
Thus in love's gentle warfare we found peace,
Where tender mercies never know surcease.

Kiss Bliss (with Euphemismus)

Document · Version 4

This six-line poem uses euphemismus in "gentle parley" (refined diplomatic conversation instead of kissing), "sweet retreat" (withdrawal into intimacy), and "tender congress" (soft assembly or meeting). These elevated expressions transform the physical act of kissing into something more ceremonial and spiritual, demonstrating how euphemismus can add grace and dignity to descriptions of intimate human moments.

When first our lips in sweet communion met,
The world grew still as morning's gentle breath,
Two hearts in tender parley, souls beget
A language known to neither life nor death.
Her mouth, a rose where honeyed secrets dwell,
Received my prayer with grace beyond compare,
While whispered vows that only lovers tell
Were sealed beneath the bower of her hair.
In that soft congress of our yearning souls,
Time paused to witness love's most sacred rite,
As passion's fire through willing vessels rolls,
Consuming darkness, kindling heaven's light.
Thus in love's gentle warfare we found peace,
Where tender mercies never know surcease.

Retry
N

Can you explain this section to me in more detail?

When first our lips in sweet communion met, The world grew still as morning's gentle breath, Two hearts in tender parley, souls beget A language known to neither life nor death.

Her mouth, a rose where honeyed secrets dwell, Received my prayer with grace beyond compare, While whispered vows that only lovers tell Were sealed beneath the bower of her hair.

In that soft congress of our yearning souls, Time paused to witness love's most sacred rite, As passion's fire through willing vessels rolls, Consuming darkness, kindling heaven's light.

Thus in love's gentle warfare we found peace, Where tender mercies never know surcease.

Edit

This sonnet uses elevated, euphemistic language to transform a simple kiss into something transcendent and sacred. Let me break down the key elements:

Opening Quatrain: "Sweet communion" presents the kiss as a religious sacrament rather than a physical act. The world becoming "still as morning's gentle breath" creates an atmosphere of reverence and natural beauty. "Tender parley" is particularly elegant euphemismus - it reimagines the meeting of lips as diplomatic negotiations between two parties. The final line suggests their kiss creates a mystical language that transcends mortality itself.

Second Quatrain: The woman's mouth becomes "a rose where honeyed secrets dwell" - a lush metaphor that combines beauty (rose), sweetness (honey), and mystery (secrets). His approach to her is described as offering "prayer," elevating his desire to an act of worship. The "bower of her hair" creates an intimate, garden-like sanctuary where their "whispered vows" are protected and consecrated.

Third Quatrain: "Soft congress" continues the euphemistic pattern - what could be described as passionate kissing becomes a gentle assembly or meeting of souls. Time itself becomes a witness to "love's most sacred rite," emphasizing the ceremonial nature of their intimacy. The "passion's fire through willing vessels" suggests they become conduits for something divine, with their union literally bringing light to darkness.

Final Couplet: "Love's gentle warfare" is perhaps the most sophisticated euphemism - it acknowledges the intensity and conquest inherent in passionate love while softening it with "gentle." The resolution finds "peace" in this paradoxical conflict, with "tender mercies" that are eternal.

The euphemistic language throughout elevates what could be a simple physical description into something spiritually profound, demonstrating how this rhetorical device can transform ordinary experience into art.

add to the wisdom of literary expression by writing a poetic quatrain about a kiss. please be sure to use an example of the word schema or figure known as euphemismus in the poem

Here's a quatrain about a kiss that employs euphemism:

A Quatrain on a Kiss

Document

The euphemism appears in "found sweet rest" and "tender congress" - gentler, more elevated expressions that replace direct descriptions of kissing and intimate connection. The phrase "sweet rest" transforms the physical act into something peaceful and restorative, while "tender congress" dignifies the meeting of lips as a formal assembly of souls rather than mere physical contact.

Editing this message will create a new conversation branch. You can switch between branches using the arrow navigation buttons.

Kiss Sonnet with Euphemismus

Document · Version 2

This Shakespearean sonnet employs euphemismus throughout to elevate and refine the description of a kiss. The euphemistic expressions include "sweet communion" (instead of kissing), "tender parley" (gentle conversation between lips), "soft congress" (intimate meeting), "love's gentle warfare" (the passionate exchange), and "love's most sacred rite" (the kiss as ceremony). These refined substitutions transform the physical act into something spiritual and transcendent, demonstrating how euphemismus can add dignity and poetic beauty to even the most common human experiences.

add to the wisdom of literary expression by writing a poetic quatrain about a kiss. please be sure to use an example of the word schema or figure known as euphemismus in the poem and add the name of the figure or schema into the title like, "kiss quatrain with euphemismus"

Kiss Quatrain with Euphemismus

Document · Version 3

This quatrain employs euphemismus through "found sweet rest" and "tender congress" - refined expressions that dignify the kiss by replacing direct description with more elevated language. "Sweet rest" transforms the physical act into something peaceful and restorative, while "tender congress" presents the meeting of lips as a formal, almost parliamentary gathering of souls rather than mere physical contact.

add to the wisdom of literary expression by writing a poem with 6 to 8 lines about a kiss. please be sure to use an example of the word schema or figure known as euphemismus in the poem and add the name of the figure or schema (paranthetically) into the title like, "kiss bliss (with euphemismus)"

Kiss Bliss (with Euphemismus)

Document · Version 4

This six-line poem uses euphemismus in "gentle parley" (refined diplomatic conversation instead of kissing), "sweet retreat" (withdrawal into intimacy), and "tender congress" (soft assembly or meeting). These elevated expressions transform the physical act of kissing into something more ceremonial and spiritual, demonstrating how euphemismus can add grace and dignity to descriptions of intimate human moments.

Can you explain this section to me in more detail?

When first our lips in sweet communion met, The world grew still as morning's gentle breath, Two hearts in tender parley, souls beget A language known to neither life nor death.

Her mouth, a rose where honeyed secrets dwell, Received my prayer with grace beyond compare, While whispered vows that only lovers tell Were sealed beneath the bower of her hair.

In that soft congress of our yearning souls, Time paused to witness love's most sacred rite, As passion's fire through willing vessels rolls, Consuming darkness, kindling heaven's light.

Thus in love's gentle warfare we found peace, Where tender mercies never know surcease.

This sonnet uses elevated, euphemistic language to transform a simple kiss into something transcendent and sacred. Let me break down the key elements:

Opening Quatrain: "Sweet communion" presents the kiss as a religious sacrament rather than a physical act. The world becoming "still as morning's gentle breath" creates an atmosphere of reverence and natural beauty. "Tender parley" is particularly elegant euphemismus - it reimagines the meeting of lips as diplomatic negotiations between two parties. The final line suggests their kiss creates a mystical language that transcends mortality itself.

Second Quatrain: The woman's mouth becomes "a rose where honeyed secrets dwell" - a lush metaphor that combines beauty (rose), sweetness (honey), and mystery (secrets). His approach to her is described as offering "prayer," elevating his desire to an act of worship. The "bower of her hair" creates an intimate, garden-like sanctuary where their "whispered vows" are protected and consecrated.

Third Quatrain: "Soft congress" continues the euphemistic pattern - what could be described as passionate kissing becomes a gentle assembly or meeting of souls. Time itself becomes a witness to "love's most sacred rite," emphasizing the ceremonial nature of their intimacy. The "passion's fire through willing vessels" suggests they become conduits for something divine, with their union literally bringing light to darkness.

Final Couplet: "Love's gentle warfare" is perhaps the most sophisticated euphemism - it acknowledges the intensity and conquest inherent in passionate love while softening it with "gentle." The resolution finds "peace" in this paradoxical conflict, with "tender mercies" that are eternal.

The euphemistic language throughout elevates what could be a simple physical description into something spiritually profound, demonstrating how this rhetorical device can transform ordinary experience into art.

add to the wisdom of literary expression by writing a poem with 6 to 8 lines about a kiss. please be sure to use an example of the word schema or figure known as euphemismus in the poem and add the name of the figure or schema (paranthetically) into the title like, "kiss bliss (with euphemismus)" also can you put and explanatory note at the bottom of the poem? and the title at the top

Kiss Bliss (with Euphemismus)

Document · Version 5

I've restructured the poem with the title at the top and added a detailed explanatory note at the bottom that identifies the specific euphemistic expressions and explains how they function to elevate the subject matter through refined substitution of language.

Kiss Bliss (with Euphemismus)

When lips in gentle parley meet,
The world dissolves in honeyed fire,
Two hearts commence their sweet retreat
From time's relentless, cold desire.
In tender congress souls unite,
While whispered prayers ignite the night.

Note: This poem employs euphemismus, a rhetorical figure that substitutes refined or elevated language for more direct expressions. "Gentle parley" transforms kissing into diplomatic conversation, "sweet retreat" dignifies intimate withdrawal from the world, and "tender congress" presents the physical meeting as a formal assembly of souls. These euphemistic substitutions elevate the kiss from mere physical contact to something ceremonial and transcendent.

add to the wisdom of literary expression by writing a poem with 6 to 8 lines about a kiss. please be sure to use an example of the word schema or figure known as euphemismus in the poem and add the name of the figure or schema (paranthetically) into the title like, "kiss bliss (with euphemismus)" also can you put and explanatory note at the bottom of the poem? and the title at the top

add to the wisdom of literary expression by writing a poem with 6 to 8 lines about a kiss. please be sure to use an example of the word schema or figure known as restrictio in the poem and add the name of the figure or schema or trope (paranthetically) into the title like, "kiss bliss (with "the name of the figure')" also can you put and explanatory note at the bottom of the poem? and the title at the beginning

add to the wisdom of literary expression by writing a poem with 6 to 8 lines about a kiss. please be sure to use an example of the word schema or figure known as euphemismus in the poem and add the name of the figure or schema (paranthetically) into the title like, "kiss poem number 2 (with euphemismus)" also can you put an explanatory note at the bottom of the poem? and the title at the top

Edit
Kiss Poem with Euphemismus
Document
Kiss Poem Number 2 (with Euphemismus) In twilight's gentle embrace we found Our lips like whispered secrets met, Where passion's language needs no sound— Just tender discourse, soft and sweet, Two souls conversing without words, In that most intim

I've crafted a poem that uses euphemismus to transform the physical act of kissing into something more refined and metaphorical. The euphemistic phrases "tender discourse" and "intimate talk" replace direct references to kissing, creating a more elevated and indirect way of describing this moment of connection. The poem explores the idea of a kiss as a form of wordless communication between two people, which allows for both beautiful imagery and the subtle rhetorical technique you requested.

Retry
N

add to the wisdom of literary expression by writing a poem with 6 to 8 lines about a kiss. please be sure to use an example of the word schema or figure known as paraenesis in the poem and add the name of the figure or schema (paranthetically) into the title like, "kiss bliss (with "the name of the figue')" also can you put and explanatory note at the bottom of the poem? and the title at the top

Edit
Kiss Poem with Paraenesis
Document
Kiss Bliss (with Paraenesis) Remember this: when lips first meet in flame, Let not your eager heart forget to pause— Breathe deep the sweetness of love's gentle game, Hold fast this moment, sacred without laws. Consider well how rare such grace ap

I've created a poem that weaves paraenesis throughout its structure, using the rhetorical figure to transform a simple kiss into a moment worthy of moral reflection and guidance. The poem becomes not just a description of romantic experience, but an exhortation to the reader about how to approach such moments—with mindfulness, reverence, and wisdom. The paraenetic voice serves as both narrator and counselor, urging the reader to treasure these fleeting but profound human experiences.

Retry
N

add to the wisdom of literary expression by writing a poem with 6 to 8 lines about a kiss. please be sure to use an example of the word schema or figure known as ominatio in the poem and add the name of the figure or schema (paranthetically) into the title like, "kiss bliss (with "the name of the figue')" also can you put and explanatory note at the bottom of the poem? and the title at the

Edit
Kiss Poem with Ominatio
Document
Kiss Bliss (with Ominatio) Your lips approach like storm clouds gathering near, And I, a fool, await what's yet to come— This kiss will be my joy, my bliss, my fear, The sweet destruction making my heart numb. I sense the lightning crackling in th

I've crafted a poem that uses ominatio to infuse a kiss with prophetic tension and foreboding. The rhetorical figure transforms what could be a simple romantic moment into something charged with anticipation of both bliss and doom. The speaker becomes almost oracular, sensing that this kiss carries within it the seeds of future joy and sorrow—a common theme in literature where moments of greatest happiness are shadowed by an awareness of their transience or consequences.

Retry
N

add to the wisdom of literary expression by writing a poem with 6 to 8 lines about a kiss. please be sure to use an example of the word schema or figure known as horismus in the poem and add the name of the figure or schema (paranthetically) into the title like, "kiss bliss (with "the name of the figue')" also can you put and explanatory note at the bottom of the poem? and the title at the

Edit
Kiss Poem with Horismus
Document
Kiss Bliss (with Horismus) What is a kiss but hearts speaking without words? A kiss: the meeting place of souls laid bare, Where breath becomes the language love prefers. A kiss is not mere touching—it's the prayer That lips inscribe on lips with

I've created a poem that uses horismus to explore the essence of a kiss through a series of definitions and explanations. Rather than simply narrating a kiss, the poem becomes an almost philosophical inquiry into what a kiss fundamentally is—transforming the physical act into various metaphysical concepts like prayer, conversation between souls, and the suspension of time itself. The horismus gives the poem a contemplative, almost scholarly tone while maintaining its romantic core.

Retry
N

add to the wisdom of literary expression by writing a poem with 6 to 8 lines about a kiss. please be sure to use an example of the word schema or figure known as systrophe in the poem and add the name of the figure or schema (paranthetically) into the title like, "kiss bliss (with "the name of the figue')" also can you put and explanatory note at the bottom of the poem? and the title at the

Edit
Kiss Poem with Systrophe
Document
Kiss Bliss (with Systrophe) Your lips, your breath, your whispered sigh— All gathered in this single space, Your warmth, your pulse, your beating heart, Your gentle hands upon my face. Each element of you draws near, Your soul, your love, your ver

I've crafted a poem that uses systrophe to create an accumulative effect, gathering multiple aspects of the beloved into a single overwhelming moment. The rhetorical figure allows the poem to build intensity through the piling up of physical and spiritual elements—lips, breath, warmth, pulse, soul—all converging in the kiss. This creates a sense of totality, as if the entire essence of the other person is being experienced simultaneously in this one intimate act.

Retry
N

add to the wisdom of literary expression by writing a poem with 6 to 8 lines about a kiss. please be sure to use an example of the word schema or figure known as diaeresis in the poem and add the name of the figure or schema (paranthetically) into the title like, "kiss bliss (with "the name of the figue')" also can you put and explanatory note at the bottom of the poem? and the title at the

Edit
Kiss Poem with Diaeresis
Document
Kiss Bliss (with Diaeresis) A kiss is not one thing but many parts: The softness first, then warmth that follows after, The breath that mingles, separate from touch, The flutter of the pulse, distinct from laughter. Each moment breaks apart like s

I've created a poem that uses diaeresis to analytically break down the experience of a kiss into its component parts and sensations. The rhetorical figure allows the poem to examine what is typically felt as one unified moment by separating it into distinct elements—softness, warmth, breath, pulse, pressure, pause. This creates an almost scientific or philosophical approach to understanding intimacy, revealing how complex and layered even our simplest human connections truly are.

Retry
N

add to the wisdom of literary expression by writing a poem with 6 to 8 lines about a kiss. please be sure to use an example of the word schema or figure known as synecdoche in the poem and add the name of the figure or schema or trope (paranthetically) into the title like, "kiss bliss (with "the name of the figue')" also can you put and explanatory note at the bottom of the poem? and the title at the beginning

Edit
Kiss Poem with Synecdoche
Document
Kiss Bliss (with Synecdoche) All eyes were watching when our lips first met, The whole world held its breath in silent pause, While every heart that ever loved could see Our kiss reflect love's universal laws. The earth itself seemed witness to ou

I've crafted a poem that uses synecdoche to expand a single kiss into a cosmic event witnessed by all of humanity and nature itself. The rhetorical figure allows the poem to suggest that this intimate moment between two people somehow connects to and represents the universal experience of love—transforming a personal kiss into something that resonates with "all eyes," "every heart," and "the whole world." The synecdoche creates both intimacy and grandeur simultaneously.

Retry
N

add to the wisdom of literary expression by writing a poem with 6 to 8 lines about a kiss. please be sure to use an example of the word schema or figure known as merismus in the poem and add the name of the figure or schema or trope (paranthetically) into the title like, "kiss bliss (with "the name of the figue')" also can you put and explanatory note at the bottom of the poem? and the title at the beginning

Edit
Kiss Poem with Merismus
Document
Kiss Bliss (with Merismus) From east to west, from dawn until the night, No kiss has ever burned with such sweet fire. From earth below to heaven's starry height, No touch has stirred such deep and pure desire. Through winter's chill and summer's

I've created a poem that uses merismus to suggest the cosmic and temporal scope of this kiss's significance. The rhetorical figure works through contrasting pairs that together represent totality—spanning all directions, all time, all creation, all seasons, and all of human life. This creates a sense that this single kiss somehow encompasses and transcends every possible comparison, making it not just intimate but universally supreme across all dimensions of existence.

Retry
N

add to the wisdom of literary expression by writing a poem with 6 to 8 lines about a kiss. please be sure to use an example of the word schema or figure known as partitio in the poem and add the name of the figure or schema or trope (paranthetically) into the title like, "kiss bliss (with "the name of the figue')" also can you put and explanatory note at the bottom of the poem? and the title at the beginning

Edit
Kiss Poem with Partitio
Document
Kiss Bliss (with Partitio) Let me speak of this kiss in three ways: First, as hunger—the starving mouth that seeks, Second, as prayer—the sacred word unspoken, Third, as answer—the silence that now speaks. Or shall I tell it thus: beginning, middl

I've crafted a poem that uses partitio to systematically organize the experience of a kiss into clear analytical categories. The rhetorical figure allows the poem to approach the kiss from multiple organizational frameworks—first as three metaphysical concepts (hunger, prayer, answer), then as three temporal stages (beginning, middle, end). This creates a sense of thoroughness and intellectual rigor, as if the speaker is determined to understand every dimension of this intimate moment through careful categorical analysis.

Retry
N

add to the wisdom of literary expression by writing a poem with 6 to 8 lines about a kiss. please be sure to use an example of the word schema or figure known as eutrepismus in the poem and add the name of the figure or schema or trope (paranthetically) into the title like, "kiss bliss (with "the name of the figue')" also can you put and explanatory note at the bottom of the poem? and the title at the beginning

Edit
Kiss Poem with Eutrepismus
Document
Kiss Bliss (with Eutrepismus) There are three kinds of kisses in this world: First, the stolen kiss—quick, guilty, wild, Second, the promised kiss—pledged and sealed, And third, the kiss that comes like answered prayer. But yours belongs to no suc

I've created a poem that uses eutrepismus to systematically catalog different types of kisses before revealing that the kiss being celebrated transcends all such categories. The rhetorical figure provides structure through enumeration while building toward the climactic revelation that this particular kiss exists beyond classification—making it both part of a systematic analysis and ultimately unique. This creates a sense of having considered all possibilities only to discover something that exceeds them all.

Retry
N

add to the wisdom of literary expression by writing a poem with 6 to 8 lines about a kiss. please be sure to use an example of the word schema or figure known as enumeratio in the poem and add the name of the figure or schema or trope (paranthetically) into the title like, "kiss bliss (with "the name of the figue')" also can you put and explanatory note at the bottom of the poem? and the title at the beginning

Edit
Kiss Poem with Enumeratio
Document
Kiss Bliss (with Enumeratio) This kiss brings fire, brings thunder, brings the rain, Brings morning light across a darkened plain, Brings healing balm to every ancient wound, Brings music where before was only pain. It brings the spring, brings ro

I've crafted a poem that uses enumeratio to create a cascading list of what this kiss brings into being. The rhetorical figure allows the poem to build through accumulation, with each "brings" clause adding another layer of metaphorical richness—from elemental forces like fire and thunder to gentle gifts like healing and starlight. This creates an almost breathless quality, as if the speaker cannot stop cataloguing all the ways this single kiss transforms their world.

Retry
N

add to the wisdom of literary expression by writing a poem with 6 to 8 lines about a kiss. please be sure to use an example of the word schema or figure known as propositio in the poem and add the name of the figure or schema or trope (paranthetically) into the title like, "kiss bliss (with "the name of the figue')" also can you put and explanatory note at the bottom of the poem? and the title at the beginning

Edit
Kiss Poem with Propositio
Document
Kiss Bliss (with Propositio) I shall prove that this kiss was destiny: First, how our paths converged from distant lands, Second, how time itself seemed to suspend, Third, how the very air grew charged and still. Consider too the way our hearts be

I've created a poem that uses propositio to transform a romantic moment into a logical argument. The rhetorical figure allows the poem to approach love with the structure of formal proof—stating a thesis (that the kiss was destined) and then methodically presenting evidence to support that claim. This creates an interesting tension between the rational, argumentative structure and the deeply emotional subject matter, suggesting that even our most profound feelings can be subjected to systematic analysis and proof.

Retry
N

add to the wisdom of literary expression by writing a poem with 6 to 8 lines about a kiss. please be sure to use an example of the word schema or figure known as restrictio in the poem and add the name of the figure or schema or trope (paranthetically) into the title like, "kiss bliss (with "the name of the figure')" also can you put and explanatory note at the bottom of the poem? and the title at the beginning

Edit
Kiss Poem with Restrictio
Document
Kiss Bliss (with Restrictio) This kiss was perfect—or would have been, If not for time that makes all moments flee. It held all love—save what remains unseen, All passion—but what lies in memory. Complete it was—except for all we lack, Pure bliss—

I've crafted a poem that uses restrictio to create a poignant tension between idealization and reality. The rhetorical figure allows the poem to simultaneously celebrate the kiss's perfection while acknowledging its inherent limitations—time's passage, memory's incompleteness, life's inevitable return to ordinary concerns. This creates a deeply human and honest portrayal of even our most transcendent moments, showing how they contain both absolute joy and the awareness of their own fragility.

Retry
N

add to the wisdom of literary expression by writing a poem with 6 to 8 lines about a kiss. please be sure to use an example of the word schema or figure known as prolepsis in the poem and add the name of the figure or schema or trope (paranthetically) into the title like, "kiss bliss (with "the name of the figure')" also can you put and explanatory note at the bottom of the poem? and the title at the beginning

Edit
Kiss Poem with Prolepsis
Document
Kiss Bliss (with Prolepsis) Before our lips have touched, I know the taste— Sweet honey mixed with morning's gentle dew. Already I can feel the wild heartbeat That hasn't quickened yet but surely will. The kiss we're yet to share has marked my sou

I've created a poem that uses prolepsis to collapse the boundaries between anticipation and fulfillment. The rhetorical figure allows the poem to treat a future kiss as if it has already happened, creating a sense of destiny and timelessness. This transforms the moment before the kiss into something almost mystical—where future experience becomes present reality through the power of love's certainty and inevitability.

Retry
N

add to the wisdom of literary expression by writing a poem with 6 to 8 lines about a kiss. please be sure to use an example of the word schema or figure known as apemantus in the poem and add the name of the figure or schema or trope (paranthetically) into the title like, "kiss bliss (with "the name of the figure')" also can you put and explanatory note at the bottom of the poem? and the title at the beginning

Edit

I'm not familiar with a rhetorical figure called "apemantus." This might be a confusion with "Apemantus," who is a character from Shakespeare's *Timon of Athens* - a cynical philosopher known for his bitter commentary.

Could you clarify what rhetorical figure you're looking for? Some possibilities that might be what you meant include:

- **Apophasis** (mentioning something by saying you won't mention it)
- **Aposiopesis** (breaking off speech suddenly)
- **Apoplanesis** (digression or evasion)
- **Apodioxis** (rejection of an opponent's argument as absurd)

Or perhaps you meant a different classical rhetorical figure? If you can provide a brief description of the effect or function you're looking for, I'd be happy to write a poem incorporating that specific technique!

Retry
N

add to the wisdom of literary expression by writing a poem with 6 to 8 lines about a kiss. please be sure to use an example of the word schema or figure known as epanodos in the poem and add the name of the figure or schema or trope (paranthetically) into the title like, "kiss bliss (with "the name of the figure')" also can you put and explanatory note at the bottom of the poem? and the title at the beginning

Edit
Kiss Poem with Epanodos
Document
Kiss Bliss (with Epanodos) Your lips find mine, and mine find yours in turn, As hearts that beat for hearts that beat the same, Where breath meets breath and souls learn how to yearn For touch that touches back without shame. The kiss that gives b

I've crafted a poem that uses epanodos to emphasize the reciprocal nature of intimate connection. The rhetorical figure creates a sense of perfect symmetry and mutual exchange, where each action is both given and received simultaneously. This captures something essential about a true kiss—that it's not one person acting upon another, but a moment of complete reciprocity where both lovers are equally active and receptive, creating a perfect circle of connection.

Retry
N

add to the wisdom of literary expression by writing a poem with 6 to 8 lines about a kiss. please be sure to use an example of the word schema or figure known as synathroesmus in the poem and add the name of the figure or schema or trope (paranthetically) into the title like, "kiss bliss (with "the name of the figure')" also can you put and explanatory note at the bottom of the poem? and the title at the beginning

Edit
Kiss Poem with Synathroesmus
Document
Kiss Bliss (with Synathroesmus) Fire, lightning, thunder, storm and rain— All gathered in the pressure of your lips. Silk, honey, wine, rose petals, morning dew, Pearl, starlight, music, incense, sacred flame, Gold, amber, crystal, diamond's brill

I've created a poem that uses synathroesmus to create an overwhelming cascade of beautiful and powerful images all compressed into the experience of a single kiss. The rhetorical figure allows the poem to suggest that this moment of intimacy somehow contains multitudes—every precious substance, every natural force, every celestial beauty all concentrated into one transformative human connection. The breathless accumulation mirrors the overwhelming nature of profound love itself.

Retry
N

add to the wisdom of literary expression by writing a poem with 6 to 8 lines about a kiss. please be sure to use an example of the word schema or figure known as congeries in the poem and add the name of the figure or schema or trope (paranthetically) into the title like, "kiss bliss (with "the name of the figure')" also can you put and explanatory note at the bottom of the poem? and the title at the beginning

Edit
Kiss Poem with Congeries
Document
Kiss Bliss (with Congeries) This kiss—this flame, this flood, this earthquake's power, This lightning strike, this thunder's rolling sound, This avalanche of sweetness crashing down, This hurricane of passion's wildest hour, This tidal wave, this

I've crafted a poem that uses congeries to pile up powerful natural disasters and phenomena as metaphors for a kiss, creating an almost overwhelming sense of force and intensity. The rhetorical figure builds through sheer accumulation of dramatic images—each one reinforcing the kiss's power while contributing to the overall effect of being completely overwhelmed by love. The final ironic twist reveals that all this apparent destruction is actually gentle tenderness, capturing the paradox of how profound love can feel both earth-shattering and completely safe.

Retry
N

add to the wisdom of literary expression by writing a poem with 6 to 8 lines about a kiss. please be sure to use an example of the word schema or figure known as epiphonema in the poem and add the name of the figure or schema or trope (paranthetically) into the title like, "kiss bliss (with "the name of the figure')" also can you put and explanatory note at the bottom of the poem? and the title at the beginning

Edit
Kiss Poem with Epiphonema
Document
Kiss Bliss (with Epiphonema) Your lips pressed soft against my waiting mouth, The world fell silent, time itself stood still, My heart forgot its rhythm, north lost south, And reason bowed before love's sovereign will. The stars above seemed dimme

I've created a poem that uses epiphonema to build toward a climactic exclamatory conclusion. The rhetorical figure allows the poem to describe all the ways this kiss transforms and overwhelms the speaker, then crystallize that entire experience into a single, memorable insight about love's "sweet tyranny." This creates a satisfying sense of completion, where all the scattered effects of the kiss are gathered into one penetrating observation about the paradoxical nature of love's power.

Retry
N

add to the wisdom of literary expression by writing a poem with 6 to 8 lines about a kiss. please be sure to use an example of the word schema or figure known as epigram in the poem and add the name of the figure or schema or trope (paranthetically) into the title like, "kiss bliss (with "the name of the figure')" also can you put and explanatory note at the bottom of the poem? and the title at the beginning

Edit
Kiss Poem with Epigram
Document
Kiss Bliss (with Epigram) We met as strangers in the morning light, Your smile, my answer to an unasked prayer. One kiss, and suddenly the world was right— What took God seven days, love did with care In just one breath between your lips and mine.

I've crafted a poem that uses epigram to create memorable, witty conclusions that capture essential truths about love and kissing. The rhetorical figure allows the poem to build through narrative description and then crystallize into sharp, paradoxical observations that play with concepts of distance, geometry, and divine creation. The epigrammatic lines provide both intellectual cleverness and profound insight, suggesting that the deepest truths about human connection can often be expressed in the most concise and surprising ways.

Retry
N

add to the wisdom of literary expression by writing a poem with 6 to 8 lines about a kiss. please be sure to use an example of the word schema or figure known as habit in the poem and add the name of the figure or schema or trope (paranthetically) into the title like, "kiss bliss (with "the name of the figure')" also can you put and explanatory note at the bottom of the poem? and the title at the beginning

Edit

I'm not familiar with "habit" as a specific classical rhetorical figure or schema. "Habit" typically refers to repeated behaviors or customs, but it's not a standard term in classical rhetoric like the other figures we've been exploring.

Could you clarify what rhetorical effect or technique you're looking for? Some possibilities that might be related to what you have in mind include:

- **Ethos** (establishing character or credibility)
- **Habitatio** (a rare figure involving dwelling on a topic)
- **Repetitio** (repetition for emphasis)
- **Conduplicatio** (repeating words or phrases to create emphasis)

Or perhaps you meant a different classical rhetorical figure? If you can describe the kind of effect or pattern you're thinking of, I'd be happy to write a poem incorporating that specific technique!

Alternatively, if you'd like me to create a poem that incorporates the concept of "habit" in terms of repeated actions or customs around kissing, I could certainly do that as well.

Retry
N

add to the wisdom of literary expression by writing a poem with 6 to 8 lines about a kiss. please be sure to use an example of the word schema or figure known as peristasis in the poem and add the name of the figure or schema or trope (paranthetically) into the title like, "kiss bliss (with "the name of the figure')" also can you put and explanatory note at the bottom of the poem? and the title at the beginning

Edit
Kiss Poem with Peristasis
Document
Kiss Bliss (with Peristasis) Though storm clouds gathered dark above our heads, Though thunder rolled and lightning split the sky, Though rain fell hard and wind shook ancient trees, Though all the world seemed bent on keeping us apart, Your kiss

I've created a poem that uses peristasis to emphasize the triumph of love over adverse circumstances. The rhetorical figure allows the poem to catalog all the obstacles and challenging conditions that surrounded this kiss, making the moment of connection even more powerful and meaningful. By detailing what the kiss overcame—natural forces, fate's opposition, worldly barriers—the poem suggests that true love possesses an almost superhuman ability to transcend any circumstance that would keep lovers apart.

Retry
N

add to the wisdom of literary expression by writing a poem with 6 to 8 lines about a kiss. please be sure to use an example of the word schema or figure known as encomium in the poem and add the name of the figure or schema or trope (paranthetically) into the title like, "kiss bliss (with "the name of the figure')" also can you put and explanatory note at the bottom of the poem? and the title at the beginning

Edit
Kiss Poem with Encomium
Document
Kiss Bliss (with Encomium) Praise be to this most sacred kiss of ours, Noble in its tenderness, pure in intent, Blessed with grace that surpasses earthly powers, Worthy of poets' songs throughout all years. Behold its virtue: healing wounded heart

I've crafted a poem that uses encomium to transform a kiss into something worthy of formal, ceremonial praise. The rhetorical figure allows the poem to adopt an elevated, almost liturgical tone that celebrates the kiss's noble qualities and excellent effects. By systematically cataloguing the kiss's virtues—its healing power, its ability to unite souls, its capacity to give mortals a taste of the divine—the poem treats this intimate human moment as something truly worthy of reverence and formal tribute.

Retry
N

add to the wisdom of literary expression by writing a poem with 6 to 8 lines about a kiss. please be sure to use an example of the word schema or figure known as taxis in the poem and add the name of the figure or schema or trope (paranthetically) into the title like, "kiss bliss (with "the name of the figure')" also can you put and explanatory note at the bottom of the poem? and the title at the beginning

Edit
Kiss Poem with Taxis
Document
Kiss Bliss (with Taxis) First came the glance that caught me unaware, Then came the smile that made my spirit soar, Next came the words we whispered soft and low, After that, the gentle touch of hands, Then came the moment when our hearts drew nea

I've created a poem that uses taxis to present the progression toward a kiss as an orderly, inevitable sequence. The rhetorical figure allows the poem to trace the natural stages of romantic connection—from first glance to final kiss—in their proper chronological order. This creates a sense that love follows its own logical pattern, with each step building naturally toward the culminating moment of the kiss, suggesting that true intimacy unfolds according to its own sacred and systematic laws.

Retry
N

add to the wisdom of literary expression by writing a poem with 6 to 8 lines about a kiss. please be sure to use an example of the word schema or figure known as epitheton in the poem and add the name of the figure or schema or trope (paranthetically) into the title like, "kiss bliss (with "the name of the figure')" also can you put and explanatory note at the bottom of the poem? and the title at the beginning

Edit
Kiss Poem with Epitheton
Document
Kiss Bliss (with Epitheton) Beneath the star-bright canopy of night, Sweet-breathing love approached with silver feet, Her rose-red lips like dawn's first gentle light Made honey-tongued promises, soft and fleet. Wine-dark passion stirred within m

I've crafted a poem that uses epitheton to create rich, ornate descriptions that elevate the kiss to mythic proportions. The rhetorical figure allows the poem to employ the kind of ceremonial, compound adjectives found in epic poetry—transforming simple elements like night, love, lips, and passion into "star-bright," "sweet-breathing," "rose-red," and "wine-dark" phenomena. This creates a timeless, almost Homeric quality that makes this intimate human moment feel worthy of epic verse.

Retry
N

add to the wisdom of literary expression by writing a poem with 6 to 8 lines about a kiss. please be sure to use an example of the word schema or figure known as a compound epithet in the poem and add the name of the figure or schema or trope (paranthetically) into the title like, "kiss bliss (with "the name of the figure')" also can you put and explanatory note at the bottom of the poem? and the title at the beginning

Edit

Kiss Poem with Compound Epithet
Document
Kiss Bliss (with Compound Epithet) The storm-crowned night bore witness to our love, As moon-bright silver lit your heart-shaped face, Your soul-deep eyes like star-kissed pools above Reflected all time's honey-sweet embrace. This fire-born moment

I've created a poem that uses compound epithets to create compressed, vivid descriptions that pack multiple meanings into hyphenated phrases. The rhetorical figure allows the poem to achieve the kind of concentrated imagery found in Anglo-Saxon poetry, where complex ideas are compressed into single descriptive units. Each compound epithet—from "storm-crowned night" to "pain-sweet kiss"—creates layers of meaning that would require much longer descriptions if expressed conventionally, giving the poem both linguistic efficiency and poetic intensity.

Retry
N

add to the wisdom of literary expression by writing a poem with 6 to 8 lines about a kiss. please be sure to use an example of the word schema or figure known as autonomasia in the poem and add the name of the figure or schema or trope (paranthetically) into the title like, "kiss bliss (with "the name of the figure')" also can you put and explanatory note at the bottom of the poem? and the title at the beginning

Edit
Kiss Poem with Autonomasia
Document
Kiss Bliss (with Autonomasia) When Venus smiled and blessed our sweet embrace, The Thunderer himself held back his storms, While Cupid, that blind archer, found his mark And pierced two hearts with one gold-feathered dart. The Muse herself would w

I've crafted a poem that uses autonomasia to elevate a simple kiss into a cosmic event witnessed and blessed by the gods themselves. The rhetorical figure allows the poem to invoke classical deities through descriptive epithets rather than direct names, creating a sense of grandeur and mythological significance. By having Venus, Jupiter, Cupid, the Muse, and Diana all take notice of this kiss, the poem suggests that true love operates on a scale worthy of divine attention and classical poetry.

Retry
N

add to the wisdom of literary expression by writing a poem with 6 to 8 lines about a kiss. please be sure to use an example of the word schema or figure known as periphrasis in the poem and add the name of the figure or schema or trope (paranthetically) into the title like, "kiss bliss (with "the name of the figure')" also can you put and explanatory note at the bottom of the poem? and the title at the beginning

Edit
Kiss Poem with Periphrasis
Document
Kiss Bliss (with Periphrasis) When rosy-fingered daughter of the morn Had fled before the chariot of night, And silver-footed huntress of the sky Cast down her gentle beams of borrowed light, We found that sweet communion of the lips That speaks w

I've created a poem that uses periphrasis to transform simple concepts into elaborate, poetic descriptions. The rhetorical figure allows the poem to avoid direct naming in favor of beautiful circumlocutions—turning dawn into "rosy-fingered daughter of the morn" and a kiss into "that sweet communion of the lips." This creates a elevated, almost Homeric style that makes familiar elements feel mythic and ceremonial, suggesting that love deserves the most ornate and indirect language poetry can provide.

Retry
N

add to the wisdom of literary expression by writing a poem with 6 to 8 lines about a kiss. please be sure to use an example of the word schema or figure known as metonymy in the poem and add the name of the figure or schema or trope (paranthetically) into the title like, "kiss bliss (with "the name of the figure')" also can you put and explanatory note at the bottom of the poem? and the title at the beginning

Edit
Kiss Poem with Metonymy
Document
Kiss Bliss (with Metonymy) The pen cannot capture what hearts felt that night, When crown and cottage stood as witnesses, As silk met wool in love's sweet democracy, While gold forgot its distance from the clay. The sword laid down before love's g

I've crafted a poem that uses metonymy to explore how love transcends social boundaries and material distinctions. The rhetorical figure allows the poem to represent larger concepts through associated objects—using "crown and cottage" to represent different social classes, "silk and wool" for economic divisions, "gold and clay" for value hierarchies. This creates a rich symbolic landscape where the kiss becomes a democratic force that unites what society typically separates, suggesting that true intimacy dissolves all artificial barriers between human beings.

Retry
N

add to the wisdom of literary expression by writing a poem with 6 to 8 lines about a kiss. please be sure to use an example of the word schema or figure known as hypotyposis in the poem and add the name of the figure or schema or trope (paranthetically) into the title like, "kiss bliss (with "the name of the figure')" also can you put and explanatory note at the bottom of the poem? and the title at the beginning

Edit
Kiss Poem with Hypotyposis
Document
Kiss Bliss (with Hypotyposis) See how her lashes flutter, dark as night, Watch how her breath comes shallow, quick, and warm, Observe the way her fingers tremble light Against my face, like leaves before a storm. Behold the crimson rising in her c

I've created a poem that uses hypotyposis to make the reader a direct witness to this intimate moment. The rhetorical figure transforms the poem into a series of visual directions—"See," "Watch," "Observe," "Behold"—that guide the reader's attention to specific, vivid details. This creates an almost cinematic effect, as if the reader is present in the scene, observing the flutter of eyelashes, the quickened breath, the trembling fingers, and the flush of emotion that accompanies this kiss.

Retry
N

add to the wisdom of literary expression by writing a poem with 6 to 8 lines about a kiss. please be sure to use an example of the word schema or figure known as enargia in the poem and add the name of the figure or schema or trope (paranthetically) into the title like, "kiss bliss (with "the name of the figure')" also can you put and explanatory note at the bottom of the poem? and the title at the beginning

Edit

Kiss Poem with Enargia
Document
Kiss Bliss (with Enargia) The jasmine's perfume mingles with her skin, Soft candlelight flickers gold across her face, I taste the wine still lingering on her lips, Feel velvet fabric whisper as we embrace. The grandfather clock ticks steady in th

I've crafted a poem that uses enargia to create a fully immersive sensory experience of this intimate moment. The rhetorical figure allows the poem to engage all five senses—the perfume of jasmine, flickering candlelight, taste of wine, touch of velvet, sounds of clock and rain—creating such vivid detail that the reader feels transported into the scene. This transforms the kiss from an abstract concept into a tangible, lived experience rich with atmospheric detail and sensory immediacy.

Retry
N

add to the wisdom of literary expression by writing a poem with 6 to 8 lines about a kiss. please be sure to use an example of the word schema or figure known as prosopographia in the poem and add the name of the figure or schema or trope (paranthetically) into the title like, "kiss bliss (with "the name of the figure')" also can you put and explanatory note at the bottom of the poem? and the title at the beginning

Edit
Kiss Poem with Prosopographia

Document
Kiss Bliss (with Prosopographia) She stands there, raven hair cascading down Past shoulders pale as morning's first snowfall, Her eyes like emeralds catching lamplight's glow, Full lips the color of wine-dark roses bloom. The ivory column of her g

I've created a poem that uses prosopographia to paint a detailed physical portrait of the beloved before the kiss. The rhetorical figure allows the poem to systematically catalog her physical beauty—from raven hair to emerald eyes to ivory neck—creating an almost sculptural sense of her presence. This detailed description builds anticipation for the kiss by making the beloved's physical beauty vividly present to the reader, transforming her from an abstract concept into a tangible, beautiful person whose kiss becomes the natural culmination of all this described loveliness.

Retry
N

add to the wisdom of literary expression by writing a poem with 6 to 8 lines about a kiss. please be sure to use an example of the word schema or figure known as prosopopoeia in the poem and add the name of the figure or schema or trope (paranthetically) into the title like, "kiss bliss (with "the name of the figure')" also can you put and explanatory note at the bottom of the poem? and the title at the beginning

Edit

Kiss Poem with Prosopopoeia
Document
Kiss Bliss (with Prosopopoeia) The moonlight whispers secrets through the trees, While gentle Wind sighs soft encouragement, And patient Night draws close her velvet sleeves To give our kiss her quiet blessing sent. The stars lean down to witness

I've crafted a poem that uses prosopopoeia to transform the natural world into conscious participants in this romantic moment. The rhetorical figure allows the poem to give human qualities to moonlight, wind, night, stars, time, and earth—making them active witnesses and supporters of this kiss. This creates a sense that love is so powerful and significant that it awakens consciousness in the very cosmos itself, with all of nature conspiring to bless and celebrate this intimate human connection.

Retry
N

add to the wisdom of literary expression by writing a poem with 6 to 8 lines about a kiss. please be sure to use an example of the word schema or figure known as characterismus in the poem and add the name of the figure or schema or trope (paranthetically) into the title like, "kiss bliss (with "the name of the figure')" also can you put and explanatory note at the bottom of the poem? and the title at the beginning

Edit
Kiss Poem with Characterismus

Document
Kiss Bliss (with Characterismus) She is the kind who laughs at morning rain, Who feeds stray cats and reads to children lost, Who gives her coat away when others freeze, Whose gentle heart bears everyone's small pain. And when this tender soul dra

I've created a poem that uses characterismus to establish the beloved's moral character and habitual kindness before describing her kiss. The rhetorical figure allows the poem to focus on her inner nature—her compassion for animals, children, and strangers—rather than physical beauty. This creates a deeper foundation for the kiss, suggesting that when someone with such a generous heart kisses you, the physical act carries within it all the goodness of their character, making it "sweet mercy pressed between two lips divine."

Retry
N

add to the wisdom of literary expression by writing a poem with 6 to 8 lines about a kiss. please be sure to use an example of the word schema or figure known as ethopoesia in the poem and add the name of the figure or schema or trope (paranthetically) into the title like, "kiss bliss (with "the name of the figure')" also can you put and explanatory note at the bottom of the poem? and the title at the beginning

Edit
Kiss Poem with Ethopoesia

Document
Kiss Bliss (with Ethopoeia) "Come closer, love," she whispers, voice like silk, "I've waited all my life for just this kiss. My heart beats wild—can you not hear it call? Oh, let me taste the sweetness of your lips, For in this moment, time itself

I've crafted a poem that uses ethopoeia to reveal the beloved's character through her own spoken words. The rhetorical figure allows the poem to show rather than tell us about her nature—through her gentle invitation, vulnerable confession, and poetic expression, we understand she possesses both tenderness and courage, passion and eloquence. Her direct speech becomes a window into her soul, making her kiss more meaningful because we've heard the heart and mind behind it speak with such beautiful honesty.

Kiss Poem Number 2 (with Euphemismus)

In twilight's gentle embrace we found

Our lips like whispered secrets met,

Where passion's language needs no sound—

Just tender discourse, soft and sweet,

Two souls conversing without words,

In that most intimate of talks,

Where hearts speak truths that can't be heard.

Note: This poem employs euphemismus in the phrase "tender discourse" and "intimate talk" to refer to kissing in refined, indirect terms. Euphemismus is a rhetorical figure that substitutes a mild, vague, or indirect expression for one that is considered too harsh, direct, or indelicate. Here, instead of directly describing the physical act of kissing, the poem uses the metaphor of conversation and discourse, elevating the kiss to a form of communication between souls.

Kiss Bliss (with Paraenesis)

Remember this: when lips first meet in flame,

Let not your eager heart forget to pause—

Breathe deep the sweetness of love's gentle game,

Hold fast this moment, sacred without laws.

Consider well how rare such grace appears,

Store up this kiss like gold against lean years,

For passion's fire, though bright, too quickly sears.

Note: This poem employs paraenesis, a rhetorical figure that offers advice, exhortation, or moral guidance. The paraenetic elements appear in the imperative phrases: "Remember this," "Let not your heart forget," "Breathe deep," "Hold fast," "Consider well," and "Store up this kiss." These direct appeals to the reader transform the poem from mere description into moral instruction about how to properly appreciate and preserve moments of love and intimacy.

Kiss Bliss (with Ominatio)

Your lips approach like storm clouds gathering near,
And I, a fool, await what's yet to come—
This kiss will be my joy, my bliss, my fear,
The sweet destruction making my heart numb.
I sense the lightning crackling in the air,
This moment holds both rapture and despair,
One kiss to lift me up, then leave me bare.

Note: This poem employs ominatio, a rhetorical figure that creates foreboding or anticipates future events, often with a sense of impending doom or significant consequence. The ominatio appears in phrases like "storm clouds gathering," "what's yet to come," "sweet destruction," and "lightning crackling in the air." The speaker anticipates that this kiss will bring both ecstasy and inevitable loss, creating dramatic tension through the prophetic sense that this moment of connection will also be a moment of vulnerability and potential heartbreak.

Kiss Bliss (with Horismus)

What is a kiss but hearts speaking without words?
A kiss: the meeting place of souls laid bare,
Where breath becomes the language love prefers.
A kiss is not mere touching—it's the prayer
That lips inscribe on lips with sacred art,
The definition of two becoming one,
A kiss: when time stops and the world departs.

Note: This poem employs horismus, a rhetorical figure that provides precise definitions or explanations of terms. The horismus appears in the definitional phrases: "What is a kiss but hearts speaking without words?" "A kiss: the meeting place of souls," "A kiss is not mere touching—it's the prayer," and "A kiss: when time stops and the world departs." Through these defining statements, the poem doesn't simply describe a kiss but attempts to capture its essential nature through a series of philosophical and poetic definitions, each building upon the last to create a fuller understanding of what a kiss truly means.

Kiss Bliss (with Systrophe)

Your lips, your breath, your whispered sigh—
All gathered in this single space,
Your warmth, your pulse, your beating heart,
Your gentle hands upon my face.
Each element of you draws near,
Your soul, your love, your very being
Converging in this kiss we share.

Note: This poem employs systrophe, a rhetorical figure that heaps together or accumulates multiple related elements, often in a series or list, to create emphasis and emotional intensity. The systrophe appears in the accumulation of the beloved's attributes: "Your lips, your breath, your whispered sigh," "Your warmth, your pulse, your beating heart," and "Your soul, your love, your very being." This piling up of elements creates a sense of overwhelming presence and intimacy, as if all aspects of the beloved are converging into this single moment of connection through the kiss.

Kiss Bliss (with Diaeresis)

A kiss is not one thing but many parts:
The softness first, then warmth that follows after,
The breath that mingles, separate from touch,
The flutter of the pulse, distinct from laughter.
Each moment breaks apart like scattered light—
The pressure, then the pause, the gentle pull,
Seven small eternities made whole.

Note: This poem employs diaeresis, a rhetorical figure that divides or separates a whole into its component parts for analysis or emphasis. The diaeresis appears in the systematic breakdown of a kiss into its constituent elements: "The softness first, then warmth," "The breath that mingles, separate from touch," "The flutter of the pulse, distinct from laughter," and "The pressure, then the pause, the gentle pull." By dissecting the kiss into these discrete sensations and moments, the poem creates a kind of anatomical or scientific examination of what is usually experienced as a unified whole, revealing the complexity hidden within a simple act of intimacy.

Kiss Bliss (with Synecdoche)

All eyes were watching when our lips first met,
The whole world held its breath in silent pause,
While every heart that ever loved could see
Our kiss reflect love's universal laws.
The earth itself seemed witness to our fire,
As countless souls throughout all time conspired
To bless this moment with their shared desire.

Note: This poem employs synecdoche, a rhetorical figure that uses a part to represent the whole or the whole to represent a part. The synecdoche appears in phrases like "All eyes were watching" (eyes representing whole people), "The whole world held its breath" (the world representing all people in it), "every heart that ever loved" (hearts representing entire persons who have experienced love), and "The earth itself seemed witness" (earth representing all of humanity and nature). These figures create a sense that this single kiss participates in something much larger—the universal human experience of love and connection.

Kiss Bliss (with Merismus)

From east to west, from dawn until the night,
No kiss has ever burned with such sweet fire.
From earth below to heaven's starry height,
No touch has stirred such deep and pure desire.
Through winter's chill and summer's blazing sun,
From first breath drawn to final heartbeat's end,
This kiss shall be the one, the only one.

Note: This poem employs merismus, a rhetorical figure that references totality by mentioning contrasting or complementary parts that together represent the whole. The merismus appears in the paired opposites: "east to west" (representing all directions/the entire world), "dawn until the night" (representing all time in a day), "earth below to heaven's starry height" (representing all of creation from lowest to highest), "winter's chill and summer's blazing sun" (representing all seasons), and "first breath drawn to final heartbeat's end" (representing an entire lifetime). These contrasting pairs work together to suggest that this kiss is supreme across all of space, time, creation, seasons, and human existence.

Kiss Bliss (with Partitio)

Let me speak of this kiss in three ways:
First, as hunger—the starving mouth that seeks,
Second, as prayer—the sacred word unspoken,
Third, as answer—the silence that now speaks.
Or shall I tell it thus: beginning, middle, end—
The lean, the touch, the sweet withdrawing sigh,
Three movements in love's perfect symphony.

Note: This poem employs partitio, a rhetorical figure that organizes discourse by dividing the subject matter into distinct parts or categories for systematic treatment. The partitio appears in two organizational schemes: first, the threefold division of the kiss as "hunger," "prayer," and "answer" (introduced by "Let me speak of this kiss in three ways"), and second, the temporal division into "beginning, middle, end" with their corresponding actions. This systematic approach creates a sense of thorough analysis, as if the speaker is carefully examining every aspect of the kiss through organized categorical thinking.

Kiss Bliss (with Eutrepismus)

There are three kinds of kisses in this world:
First, the stolen kiss—quick, guilty, wild,
Second, the promised kiss—pledged and sealed,
And third, the kiss that comes like answered prayer.
But yours belongs to no such common class—
It is the fourth kind, rarest of them all:
The kiss that makes all other kisses past.

Note: This poem employs eutrepismus, a rhetorical figure that enumerates different possibilities or alternatives in an orderly fashion, often to demonstrate completeness of consideration before arriving at a conclusion. The eutrepismus appears in the systematic listing of "three kinds of kisses"—the stolen, the promised, and the prayer-like kiss—each numbered and defined. However, the poem then transcends this classification by declaring that the subject kiss represents a "fourth kind" that surpasses all categories. This creates both systematic thoroughness and ultimate uniqueness, suggesting that while we can categorize most experiences, true love transcends all classification.

Kiss Bliss (with Enumeratio)

This kiss brings fire, brings thunder, brings the rain,
Brings morning light across a darkened plain,
Brings healing balm to every ancient wound,
Brings music where before was only pain.
It brings the spring, brings roses into bloom,
Brings starlight dancing through a midnight room,
Brings all the world's sweet mercy in one breath.

Note: This poem employs enumeratio, a rhetorical figure that consists of a detailed listing or enumeration of various particulars, often in a rapid, accumulative sequence. The enumeratio appears in the sustained series of parallel clauses beginning with "brings"—fire, thunder, rain, morning light, healing balm, music, spring, roses, starlight, and mercy. Each item in this enumeration represents a different metaphorical gift or effect of the kiss, creating an overwhelming sense of abundance and transformation. The repetitive structure builds momentum and emphasizes the kiss's power to bring about multiple forms of renewal and joy.

Kiss Bliss (with Propositio)

I shall prove that this kiss was destiny:
First, how our paths converged from distant lands,
Second, how time itself seemed to suspend,
Third, how the very air grew charged and still.
Consider too the way our hearts beat true
In rhythm matched before our lips could meet—
All evidence that fate had made this kiss.

Note: This poem employs propositio, a rhetorical figure that states what the speaker intends to prove or demonstrate, typically followed by the supporting arguments or evidence. The propositio appears in the opening declaration: "I shall prove that this kiss was destiny." The poem then proceeds to offer numbered evidence ("First... Second... Third...") and additional supporting observations ("Consider too...") to substantiate this central claim. This creates a logical, almost legal structure where the kiss becomes the subject of a reasoned argument, transforming romantic experience into a case to be proven through accumulated evidence.

Kiss Bliss (with Restrictio)

This kiss was perfect—or would have been,
If not for time that makes all moments flee.
It held all love—save what remains unseen,
All passion—but what lies in memory.
Complete it was—except for all we lack,
Pure bliss—though shadows linger at the edge.
All heaven—yet we still must journey back.

Note: This poem employs restrictio, a rhetorical figure that first makes a broad statement and then immediately narrows, qualifies, or restricts that statement with exceptions or limitations. The restrictio appears in each couplet through the pattern of assertion followed by qualification: "perfect—or would have been, if not for...," "held all love—save what remains...," "All passion—but what lies...," "Complete it was—except for...," "Pure bliss—though shadows...," and "All heaven—yet we still must..." This creates a bittersweet tension where each celebration of the kiss's perfection is immediately tempered by an acknowledgment of its limitations or transience.

Kiss Bliss (with Prolepsis)

Before our lips have touched, I know the taste—
Sweet honey mixed with morning's gentle dew.
Already I can feel the wild heartbeat
That hasn't quickened yet but surely will.
The kiss we're yet to share has marked my soul,
Its memory burns bright though not yet born,
Tomorrow's tenderness alive today.

Note: This poem employs prolepsis, a rhetorical figure that anticipates and describes future events as if they have already occurred, or treats something that will happen as if it is happening now. The prolepsis appears in phrases like "Before our lips have touched, I know the taste," "Already I can feel the wild heartbeat / That hasn't quickened yet," "The kiss we're yet to share has marked my soul," and "Tomorrow's tenderness alive today." This creates a sense of inevitability and timelessness, suggesting that true love transcends chronological time—the future kiss is so certain and powerful that it already exists in the present moment of anticipation.

Kiss Bliss (with Epanodos)

Your lips find mine, and mine find yours in turn,
As hearts that beat for hearts that beat the same,
Where breath meets breath and souls learn how to yearn
For touch that touches back without shame.
The kiss that gives becomes the kiss received,
Love loving love in perfect symmetry,
Two made as one, yet one that lives as two.

Note: This poem employs epanodos, a rhetorical figure that creates reciprocal or mirrored relationships between elements, often through chiastic or reversible structures. The epanodos appears in phrases like "Your lips find mine, and mine find yours," "hearts that beat for hearts that beat," "breath meets breath," "touch that touches back," "The kiss that gives becomes the kiss received," and "Love loving love." This creates a sense of perfect reciprocity and mutual exchange, emphasizing how true intimacy involves not just giving or receiving, but a simultaneous interchange where each partner becomes both subject and object of the same loving action.

Kiss Bliss (with Synathroesmus)

Fire, lightning, thunder, storm and rain—
All gathered in the pressure of your lips.
Silk, honey, wine, rose petals, morning dew,
Pearl, starlight, music, incense, sacred flame,
Gold, amber, crystal, diamond's brilliant gleam,
Heaven's choir, angel wings, and paradise—
All crowded in this single kiss we share.

Note: This poem employs synathroesmus, a rhetorical figure that heaps together or accumulates many words, phrases, or concepts in rapid succession to create an overwhelming effect of abundance or intensity. The synathroesmus appears in the dense clustering of images: "Fire, lightning, thunder, storm and rain," followed by "Silk, honey, wine, rose petals, morning dew," then "Pearl, starlight, music, incense, sacred flame," and finally "Gold, amber, crystal, diamond's brilliant gleam" and "Heaven's choir, angel wings, and paradise." This creates an almost breathless accumulation of beautiful and powerful things, suggesting that this single kiss somehow contains within it all the most precious and intense experiences possible.

Kiss Bliss (with Congeries)

This kiss—this flame, this flood, this earthquake's
power,
This lightning strike, this thunder's rolling sound,
This avalanche of sweetness crashing down,
This hurricane of passion's wildest hour,
This tidal wave, this blazing meteor's flight,
This volcano's eruption burning bright—
Destroys me with its gentle, tender might.

Note: This poem employs congeries, a rhetorical figure that accumulates a heap or mass of words or phrases with similar meanings to create emphasis through sheer abundance. The congeries appears in the sustained series of powerful natural phenomena used to describe the kiss: "flame," "flood," "earthquake's power," "lightning strike," "thunder's rolling sound," "avalanche," "hurricane," "tidal wave," "blazing meteor's flight," and "volcano's eruption." Each metaphor reinforces the kiss's overwhelming force while building cumulative intensity. The final line creates ironic contrast by revealing that all this destructive natural power is actually "gentle, tender might," emphasizing how love can be simultaneously overwhelming and nurturing.

Kiss Bliss (with Epiphonema)

Your lips pressed soft against my waiting mouth,
The world fell silent, time itself stood still,
My heart forgot its rhythm, north lost south,
And reason bowed before love's sovereign will.
The stars above seemed dimmer than your eyes,
The earth beneath felt less solid than your touch—
O what sweet tyranny lies in a kiss!

Note: This poem employs epiphonema, a rhetorical figure that concludes a passage with a short, pithy, and often exclamatory statement that summarizes or reflects upon what has been described. The epiphonema appears in the final line: "O what sweet tyranny lies in a kiss!" This exclamatory conclusion draws together all the preceding images of disruption and transformation—the silenced world, stopped time, forgotten heartbeat, lost directions, dimmed stars, and unstable earth—into a single, memorable reflection that captures the essence of how love's power overwhelms us completely, yet we willingly submit to its beautiful dominion.

Kiss Bliss (with Epigram)

We met as strangers in the morning light,
Your smile, my answer to an unasked prayer.
One kiss, and suddenly the world was right—
What took God seven days, love did with care
In just one breath between your lips and mine.
A kiss: the shortest distance between souls.
Love's geometry defies all earthly rules.

Note: This poem employs epigram, a rhetorical figure that presents a concise, witty, and often paradoxical statement that encapsulates a larger truth or observation. The epigrammatic elements appear in the final two lines: "A kiss: the shortest distance between souls" and "Love's geometry defies all earthly rules." These condensed, memorable statements distill the entire experience of the kiss into pithy observations that play with mathematical and geometric concepts to express profound truths about love's power to connect people and transcend ordinary logic. The epigrams provide both intellectual wit and emotional insight.

Kiss Bliss (with Peristasis)

Though storm clouds gathered dark above our heads,
Though thunder rolled and lightning split the sky,
Though rain fell hard and wind shook ancient trees,
Though all the world seemed bent on keeping us apart,
Your kiss broke through each barrier fate had made—
No tempest fierce enough to stop true love,
No circumstance too strong for hearts that burn.

Note: This poem employs peristasis, a rhetorical figure that describes or enumerates the circumstances, conditions, or obstacles surrounding an event, often to emphasize how remarkable it is that the event occurred despite these challenges. The peristasis appears in the sustained series of "Though" clauses that detail the adverse conditions: storm clouds, thunder, lightning, rain, wind, and the world's opposition. By cataloguing all these obstacles, the poem emphasizes the power and determination of love to overcome any circumstance. The kiss becomes more meaningful precisely because it happened despite—or perhaps because of—the dramatic challenges that surrounded it.

Kiss Bliss (with Encomium)

Praise be to this most sacred kiss of ours,
Noble in its tenderness, pure in intent,
Blessed with grace that surpasses earthly powers,
Worthy of poets' songs throughout all years.
Behold its virtue: healing wounded hearts,
Its excellence: uniting separated souls,
Its glory: making mortals taste of heaven.

Note: This poem employs encomium, a rhetorical figure that offers formal praise or tribute to a person, thing, or concept by cataloguing its virtues and excellent qualities. The encomium appears in the systematic praise of the kiss through elevated language: "most sacred," "Noble," "pure," "Blessed with grace," "surpasses earthly powers," and "Worthy of poets' songs." The poem then formally lists the kiss's virtues ("healing wounded hearts"), excellence ("uniting separated souls"), and glory ("making mortals taste of heaven"). This creates a ceremonial, almost religious tone that elevates the simple human act of kissing to something deserving of formal celebration and reverence.

Kiss Bliss (with Taxis)

First came the glance that caught me unaware,
Then came the smile that made my spirit soar,
Next came the words we whispered soft and low,
After that, the gentle touch of hands,
Then came the moment when our hearts drew near,
Finally, the kiss that sealed our fate complete,
Each step arranged by love's own sacred law.

Note: This poem employs taxis, a rhetorical figure that presents events or elements in their proper chronological or logical order, creating a sense of systematic arrangement and progression. The taxis appears in the carefully ordered sequence marked by temporal connectors: "First," "Then," "Next," "After that," "Then," and "Finally." This creates a sense of inevitable progression from initial attraction through various stages of connection to the culminating kiss. The orderly arrangement suggests that love follows its own natural sequence—that there is a proper order to intimacy that unfolds according to "love's own sacred law."

Kiss Bliss (with Epitheton)

Beneath the star-bright canopy of night,
Sweet-breathing love approached with silver feet,
Her rose-red lips like dawn's first gentle light
Made honey-tongued promises, soft and fleet.
Wine-dark passion stirred within my soul
As fire-kissed moments blazed beyond control,
This golden-hearted kiss made two hearts whole.

Note: This poem employs epitheton, a rhetorical figure that uses descriptive adjectives or phrases attached to nouns to create vivid, often formulaic characterizations. The epitheton appears in the compound adjectives and descriptive phrases: "star-bright canopy," "Sweet-breathing love," "silver feet," "rose-red lips," "honey-tongued promises," "Wine-dark passion," "fire-kissed moments," and "golden-hearted kiss." These epithets create a rich, almost Homeric quality to the language, elevating the simple act of kissing through ornate, ceremonial description that makes the experience feel mythic and timeless.

Kiss Bliss (with Compound Epithet)

The storm-crowned night bore witness to our love,
As moon-bright silver lit your heart-shaped face,
Your soul-deep eyes like star-kissed pools above
Reflected all time's honey-sweet embrace.
This fire-born moment, thunder-blessed and pure,
Made earth-bound mortals taste the sky-born sure—
One breath-stopped kiss, both pain-sweet and secure.

Note: This poem employs compound epithets, rhetorical figures that combine two or more words (usually with hyphens) to create vivid, concentrated descriptions that would require longer phrases to express otherwise. The compound epithets appear throughout: "storm-crowned night," "moon-bright silver," "heart-shaped face," "soul-deep eyes," "star-kissed pools," "honey-sweet embrace," "fire-born moment," "thunder-blessed," "earth-bound mortals," "sky-born," "breath-stopped kiss," and "pain-sweet." These compressed, hyphenated descriptions create an intense, almost Anglo-Saxon poetic quality that packs multiple meanings into single descriptive units, making the language feel both ancient and immediate.

Kiss Bliss (with Autonomasia)

When Venus smiled and blessed our sweet embrace,
The Thunderer himself held back his storms,
While Cupid, that blind archer, found his mark
And pierced two hearts with one gold-feathered dart.
The Muse herself would weep to sing such bliss,
As Dawn's bright sister watched from silver throne—
One kiss to make the very gods take notice.

Note: This poem employs autonomasia, a rhetorical figure that substitutes a descriptive phrase or epithet for a proper name, or uses a proper name to represent a general class. The autonomasia appears in several forms: "Venus" (representing love/beauty), "the Thunderer" (for Zeus/Jupiter), "that blind archer" (for Cupid), "The Muse" (representing poetic inspiration), and "Dawn's bright sister" (for the Moon/Diana). Rather than using common nouns or direct names, the poem invokes these classical figures through descriptive titles and epithets, creating a sense that this kiss is significant enough to involve the pantheon of gods and mythological figures in its celebration.

Kiss Bliss (with Periphrasis)

When rosy-fingered daughter of the morn
Had fled before the chariot of night,
And silver-footed huntress of the sky
Cast down her gentle beams of borrowed light,
We found that sweet communion of the lips
That speaks without the need for words or sound—
Love's silent language, tender and profound.

Note: This poem employs periphrasis, a rhetorical figure that uses a roundabout or indirect way of expressing something, often through elaborate description rather than direct naming. The periphrasis appears in phrases like "rosy-fingered daughter of the morn" (for Dawn/Aurora), "the chariot of night" (for nightfall/darkness), "silver-footed huntress of the sky" (for the Moon/Diana), "her gentle beams of borrowed light" (for moonlight), "that sweet communion of the lips" (for a kiss), and "Love's silent language" (also for kissing). Rather than using simple, direct terms, the poem creates elaborate, poetic circumlocutions that elevate the language and make even familiar concepts feel fresh and ceremonial.

Kiss Bliss (with Metonymy)

The pen cannot capture what hearts felt that night,
When crown and cottage stood as witnesses,
As silk met wool in love's sweet democracy,
While gold forgot its distance from the clay.
The sword laid down before love's gentle reign,
And ivory whispered secrets to the earth—
One kiss to bridge what birth had set apart.

Note: This poem employs metonymy, a rhetorical figure that substitutes the name of something with the name of something else closely associated with it. The metonymies appear throughout: "The pen" (representing writing/literature), "crown and cottage" (representing royalty and commoners), "silk met wool" (representing upper and lower classes through their clothing), "gold" and "clay" (representing precious and common materials/people), "The sword" (representing warfare/conflict), and "ivory" and "earth" (representing luxury and the common ground). These substitutions create a sense that this kiss transcends all social boundaries and material distinctions, uniting opposites through the power of love.

Kiss Bliss (with Hypotyposis)

See how her lashes flutter, dark as night,
Watch how her breath comes shallow, quick, and warm,
Observe the way her fingers tremble light
Against my face, like leaves before a storm.
Behold the crimson rising in her cheeks,
The silver tear that catches morning's gleam—
Here, in this moment, love itself now speaks.

Note: This poem employs hypotyposis, a rhetorical figure that presents a vivid, detailed description designed to place the scene clearly before the reader's eyes, as if they were witnessing it directly. The hypotyposis appears in the series of imperative verbs that direct the reader's attention: "See how her lashes flutter," "Watch how her breath comes," "Observe the way her fingers tremble," and "Behold the crimson rising." These commands create an almost cinematic quality, guiding the reader's eye to specific visual and tactile details that make the scene immediate and present. The technique transforms the reader from passive audience into active witness of this intimate moment.

Kiss Bliss (with Enargia)

The jasmine's perfume mingles with her skin,
Soft candlelight flickers gold across her face,
I taste the wine still lingering on her lips,
Feel velvet fabric whisper as we embrace.
The grandfather clock ticks steady in the hall,
Rain patters gentle rhythms on the glass—
Each sense alive to capture love's sweet call.

Note: This poem employs enargia, a rhetorical figure that creates vivid, detailed description appealing to multiple senses to make a scene come alive with striking clarity and immediacy. The enargia appears in the rich sensory details: smell ("jasmine's perfume"), sight ("candlelight flickers gold"), taste ("wine still lingering"), touch ("velvet fabric whisper"), and sound ("grandfather clock ticks," "rain patters gentle rhythms"). By engaging all the senses simultaneously, the poem creates an immersive experience that makes the reader feel physically present in this intimate moment, transforming abstract emotion into concrete, lived experience.

Kiss Bliss (with Prosopographia)

She stands there, raven hair cascading down
Past shoulders pale as morning's first snowfall,
Her eyes like emeralds catching lamplight's glow,
Full lips the color of wine-dark roses bloom.
The ivory column of her graceful neck
Curves toward me, while her slender fingers trace
My cheek before our destined kiss takes place.

Note: This poem employs prosopographia, a rhetorical figure that provides detailed physical description of a person's appearance, often focusing on specific bodily features to create a vivid portrait. The prosopographia appears in the systematic catalog of the beloved's physical attributes: "raven hair cascading down," "shoulders pale as morning's first snowfall," "eyes like emeralds catching lamplight's glow," "Full lips the color of wine-dark roses," "The ivory column of her graceful neck," and "slender fingers." This detailed physical portraiture creates an almost sculptural sense of the beloved's beauty, making her presence tangible and immediate before the culminating moment of the kiss.

Kiss Bliss (with Prosopopoeia)

The moonlight whispers secrets through the trees,
While gentle Wind sighs soft encouragement,
And patient Night draws close her velvet sleeves
To give our kiss her quiet blessing sent.
The stars lean down to witness love's sweet art,
As Time himself holds still his racing heart,
And Earth below hums low her ancient song.

Note: This poem employs prosopopoeia, a rhetorical figure that gives human characteristics, actions, or speech to non-human things, abstract concepts, or inanimate objects. The prosopopoeia appears throughout: "moonlight whispers secrets," "Wind sighs soft encouragement," "Night draws close her velvet sleeves" and gives blessing, "stars lean down to witness," "Time himself holds still his racing heart," and "Earth below hums low her ancient song." By personifying these natural forces and abstract concepts, the poem creates a sense that the entire universe is consciously participating in and blessing this moment of human love.

Kiss Bliss (with Characterismus)

She is the kind who laughs at morning rain,
Who feeds stray cats and reads to children lost,
Who gives her coat away when others freeze,
Whose gentle heart bears everyone's small pain.
And when this tender soul draws close to mine,
Her kiss tastes of all the good she's done—
Sweet mercy pressed between two lips divine.

Note: This poem employs characterismus, a rhetorical figure that describes a person's character, moral qualities, habits, or typical behaviors rather than focusing on physical appearance. The characterismus appears in the catalog of the beloved's character traits and actions: "laughs at morning rain," "feeds stray cats," "reads to children lost," "gives her coat away when others freeze," and "gentle heart bears everyone's small pain." By establishing her moral character first—her kindness, generosity, and compassion—the poem suggests that her kiss carries within it all these beautiful qualities, making it not just a physical act but an expression of her entire loving nature.

Kiss Bliss (with Ethopoeia)

"Come closer, love," she whispers, voice like silk,
"I've waited all my life for just this kiss.
My heart beats wild—can you not hear it call?
Oh, let me taste the sweetness of your lips,
For in this moment, time itself stands still,
And all the world fades soft around us here."
Her words reveal a soul both brave and tender.

Note: This poem employs ethopoeia, a rhetorical figure that reveals character through the representation of speech, showing what kind of person someone is through their actual words and manner of speaking. The ethopoeia appears in the direct quotation that captures the beloved's voice: her gentle invitation ("Come closer, love"), her vulnerability ("I've waited all my life"), her passionate honesty ("My heart beats wild"), her romantic yearning ("let me taste the sweetness"), and her poetic sensibility ("time itself stands still"). Through these quoted words, we understand her character as someone who is simultaneously tender and bold, romantic yet direct, passionate but eloquent.

add to the wisdom of literary expression by writing a poem with 6 to 8 lines about a kiss. please be sure to use an example of the word schema or figure known as ethopoesia in the poem and add the name of the figure or schema or trope (paranthetically) into the title like, "kiss bliss (with "the name of the figure')" also can you put and explanatory note at the bottom of the poem? and the title at the beginningpage 176 ethopoeia

Made in the USA
Monee, IL
04 November 2026

67391554R00144